# Synthesis Lectures on Engineering, Science, and Technology

The focus of this series is general topics, and applications about, and for, engineers and scientists on a wide array of applications, methods, and advances. Most titles cover subjects such as professional development, education, and study skills, as well as basic introductory undergraduate material and other topics appropriate for a broader and less technical audience.

Arthur David Snider

# From STEM to STEAM

## A Study of Art from the Perspective of Science, Technology, Engineering, and Mathematics

Springer

Arthur David Snider
Department of Electrical Engineering
University of South Florida
Tampa, FL, USA

ISSN 2690-0300 ISSN 2690-0327 (electronic)
Synthesis Lectures on Engineering, Science, and Technology
ISBN 978-3-031-57315-6 ISBN 978-3-031-57316-3 (eBook)
https://doi.org/10.1007/978-3-031-57316-3



This Springer imprint is published by the registered company Springer Nature Switzerland AG
The registered company address is: Gewerbestrasse 11, 6330 Cham, Switzerland

*The book is dedicated to my friends and colleagues Joshua Nicodemus and Debby and Rick Levinson.*

*Joshua was my graduate assistant during the spawning of the new elective course Light and the Arts and his indispensable insights, drawings, and criticisms are reflected throughout the volume.*

*Debby Levinson earned her doctorate under my direction in 1988. One evening back then she and her husband Rick were visiting my home when I bemoaned my ignorance of fine art. I voiced that I wished there were a "catalog" book of famous paintings, because I recognized a lot of paintings but didn't know their titles or the artists. I tried to describe one of my favorites, which I could only conjecture was called "Venus on the Half Shell". It depicted Venus standing on what appeared to be a giant oyster shell, floating on the sea and being blown to land by some angels. Debby and Rick exchanged a glance, and simultaneously said "Janson!". Debby then said "Christmas." (Rick sagely informed me that I had correctly surmised the name of Botticelli's masterpiece. Fig. 1.) The next Christmas they gifted me with a copy of their textbook as undergraduates, the renowned*

*Janson's History of Art.*[1] *It jump-started my transition from art curmudgeon to art enthusiast. Venus on the Half Shell was my first working title for the present volume.*

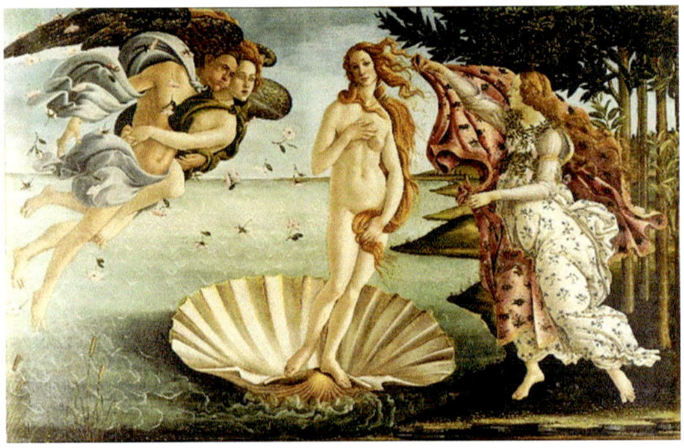

**Fig. 1** *Birth of Venus* (**Botticelli**) By Sandro Botticelli—Adjusted levels from File:Sandro Botticelli—La nascita di Venere—Google Art Project.jpg, originally from Google Art Project. Compression Photoshop level 9. Public Domain, https://commons.wikimedia.org/w/index.php?curid=22507491

[1] Janson, H. W. (1986) *History of Art*, third edition, Harry N. Abrams, New York.

# Preface

In 2004 I received a grant from the National Science Foundation that enabled me to develop an elective course in art appreciation for science, technology, engineering, and mathematics students. This book is one of the outgrowths of that project.

Typical STEMs, like my younger self, know little about art and are intimidated by liberal art classes, to the extent that we pretend to be disdainful of the curriculum while secretly dreading its peril (low grades). The "hook" that distinguishes my course and my book is the vantage point I take in approaching the subject. I create a different kind of art experience, where STEM has the upper hand. I get the students to focus on Michaelangelo's David, not by rhapsodizing about its aesthetic qualities, but by challenging its mechanical stability. I show them how their vector analysis simplifies and extends the ponderous pre-calculus rules for perspective employed by Leonardo and Raphael. I build upon their recognition of color as a simple manifestation of wavelength and go on to describe the heavy-handed mechanisms for *detecting* color employed by the eye, which enables its manipulation by artisans. These revelations whet their appetite for more examples. And they're hooked.

In short, my main objective in selling art is to get STEMs to *look at it*. Then it will sell itself. I am not above appealing to their prurient interests, their sense of humor, and sometimes their gullibility in pursuit of this objective.

## Structure of the Book

Chapter 1 is simply an introduction—getting acquainted. In Chap. 2 we tell the classical story (perhaps fanciful) of Michaelangelo's challenge in creating his whopping masterpiece. This motivates a scrutinization of David to detect his fragile points, and a crude stress analysis of them. Since our guileless STEM has now invested some time in perusing David, it is not hard to direct his/her attention to the aesthetic choices Michaelangelo made and compare them with those of Donatello and Bellini; and then to compare his Moses with others', and finally to survey the contributions of the rest of his sculptor lodge-brothers. The previously unheralded role of Vasari, recently revealed by bioengineer Maurizio Seracini's quest for the missing da Vinci painting, makes a fascinating narrative for STEM specialists.

Chapter 3 examines the discovery of perspective by Filippo Brunelleschi and reformulates it in the STEM's vernacular of vector analysis. This leads to a much cleaner expostulation, even permitting a transparent derivation of the "$1/n$" rule for drawing pavements. But possibly a bigger payoff is the inspirational (and amusing) chronicle of Brunelleschi's engineering career. The quantum leap in realistic drawing that perspective enabled, during the Italian Renaissance, motivates a closer look at the works of da Vinci and Raphael. And of course, no STEM needs prodding to appreciate the fanciful conjurings of Escher, Penrose, Necker, and Hogarth.

Chapters 4 and 5 examine color in all its aspects. After a refresher on its role in the spectral decomposition of light, Chap. 4 turns to the eye's mechanisms for detecting color, stressing the analogies of its cones to nonideal signal-processing transducers. The temptation to exploit their imperfections leads to the additive theory of color, the revelation of nonspectral colors and metamers, lighting applications, and the C.I.E. chromaticity diagram.

Chapter 5 turns to the technological aspects of how artists manipulate color, including the subtractive theory and the Munsell tree. Color juxtaposition prompts a discussion of Seurat and Lichtenstein, and color consciousness in general inspires the examination of Modigliani, Soutine, van Gogh, and Cezanne. Some historians have attributed the latter's work with premonitions of *binocular* perspective, prompting a brief reevaluation of the deliberations of Chap. 3.

## Choice of Subject Matter

The selection of artworks is unapologetically personal. I'm out to create art fans, not scholars, and these works have a proven track record; they converted a wizened old technogeek (yours truly) into a devoted enthusiast—devoted enough to write a book.

Any STEM professor has the qualifications to teach from the book, because the requisite art lore is provided (over 280 images). My text is addressed, not to the aficionado, but to the STEM practitioner coming in at ground zero in art schooling. It is, intentionally, not profound.

Tampa, USA Arthur David Snider

## Creative Commons License Links

Most of the reproductions in the text carry one of the following Creative Commons licenses. For convenience, I have gathered in one place the URLs that describe each particular license:



## Other Books by the Author

1. *Introduction to Vector Analysis*, (with H. Davis), Hawkes Learning; 7th ed., 1994; Japanese edition, 1983; Spanish edition, 1992. (ISBN 978-0-697-16099-7)
2. *Fundamentals of Complex Analysis for Mathematicians, Scientists, and Engineers* (with E. B. Saff), Pearson, 3rd ed. 2003. (ISBN 0-13-907874-6)
3. *Fundamentals of Differential Equations and Boundary Value Problems* (with K. Nagle and E. B. Saff), Pearson, 9th ed. 2017. (ISBN 13: 978-0-321-74774-7)
4. *Partial Differential Equations—Sources and Solutions*, Prentice-Hall, 1999; Dover Pub., 2006. (ISBN 0-486-45340-5)
5. *Fundamentals of Linear Algebra with Applications*, (with E. B. Saff), Springer Nature Switzerland AG, 2024 (to appear).
6. *Random Processes for Engineers: A Primer*, CRC Press (Taylor-Francis), 2017 (ISBN 13: 978-1-4987-9903-4).
7. *Basics of Optimization*, Springer Synthesis Lectures on Math and Statistics, 2023 (ISBN 978-3-031-29218-7).

**Acknowledgements** This work was supported in part by National Science Foundation Grant No. 0343740.

# Contents

# 1 Introduction

You're an engineer/scientist, and you want to learn how to appreciate art? I was in your shoes a few years ago, and I understand. You've been captivated by the sensuality, the wit, pageantry, and illusion of some of the works you have seen —but can't identify them. Any of these look familiar (Figs. 1.1, 1.2, 1.3 and 1.4)?[1]

Some of you may have even tried to take an art history class. Bless your heart. So you know what it feels like to be out of your depth. All of your artsy classmates possess a gift for drawing and for the jargon of art appreciation. You don't get it when they go on about "painterly qualities". You can't talk the talk, you can't walk the walk, you can't draw without a straightedge and compass and CAD software. (My Wellesley College girlfriends even had a game called "Botticelli," whose only recognizable merit was to humiliate geeks like us.)

Well, this book was created for a different kind of art class, where the engineer has the upper hand. They say "hue", you retort with "wavelength". They say "vanishing point", you say "asymptote." They say "brushstroke segmentation", you counterpunch with "edge detection". Not to mention how you beat your chest while expostulating "restoration techniques" and "overpainting detectors". Hah!

[1] Creative Commons License Links for all images are listed in the Preface.



A. D. Snider, *From STEM to STEAM*, Synthesis Lectures on Engineering, Science, and Technology, https://doi.org/10.1007/978-3-031-57316-3_1

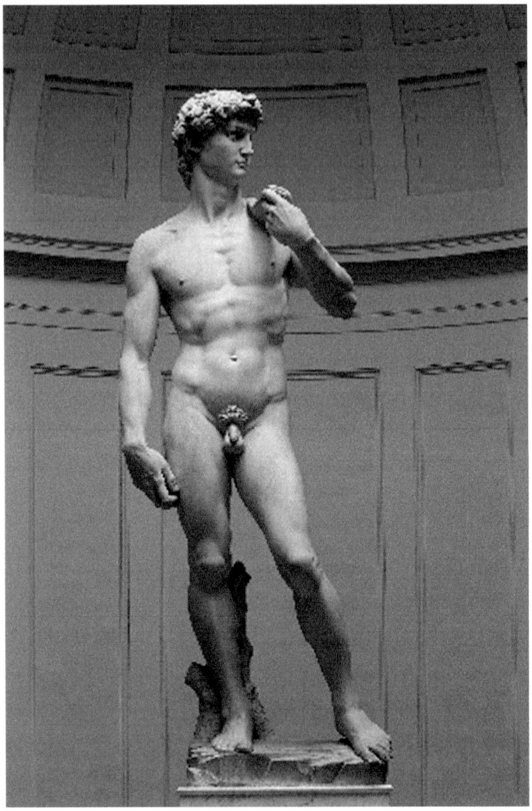

**Fig. 1.1** *David* **(Michelangelo).** By Jörg Bittner Unna—own work, CC BY 3.0, https://commons.wikimedia.org/w/index.php?curid=56633968. c. 1504; Galleria dell'Accademia, Florence

Ready? We're going to start with Michelangelo's sculpture of David. You're going to give him a safety inspection. Then you're going to trace Leonardo's Last Supper like a draftsman, testing his adherence to perspective theory. You'll see how a color TV uses subpixels of only three different colors to convince the eye it's seeing the full splendor of a Rose Bowl parade or a Cezanne masterpiece. And as a byproduct, unless I miss my guess you will painlessly acquire an appreciation for fine art. Which is why you bought my book.

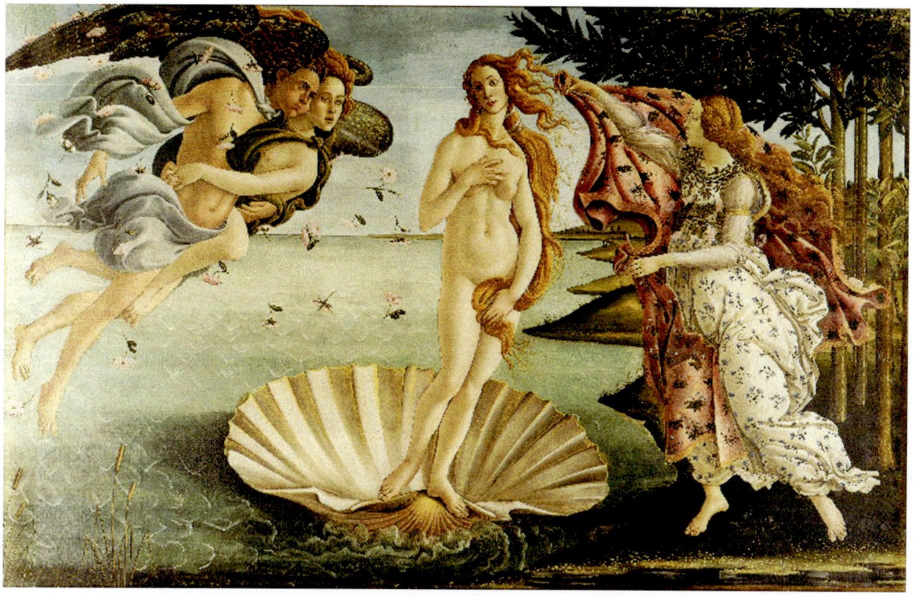

**Fig. 1.2** *Birth of venus* **(Botticelli)** By Sandro Botticelli—Adjusted levels from file:Sandro Botticelli—La nascita di Venere—Google art project.jpg, originally from google art project. Compression photoshop level 9., public domain, https://commons.wikimedia.org/w/index.php?curid=22507491. c. 1485; Uffizi Gallery, Florence

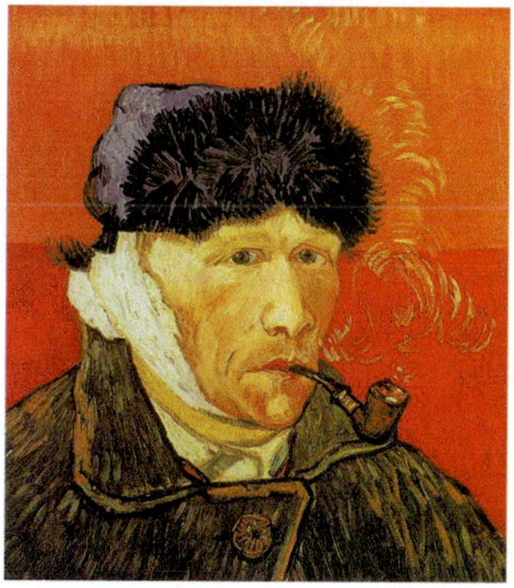

**Fig. 1.3** *Selbstbildnis_mit_verbundenem* (**Van_Gogh**). By Vincent van Gogh—repro from artbook, public domain, https://commons.wikimedia.org/w/index.php?curid=9485706. {{PD-US-expired}} (1889; Kunsthaus, Zürich)

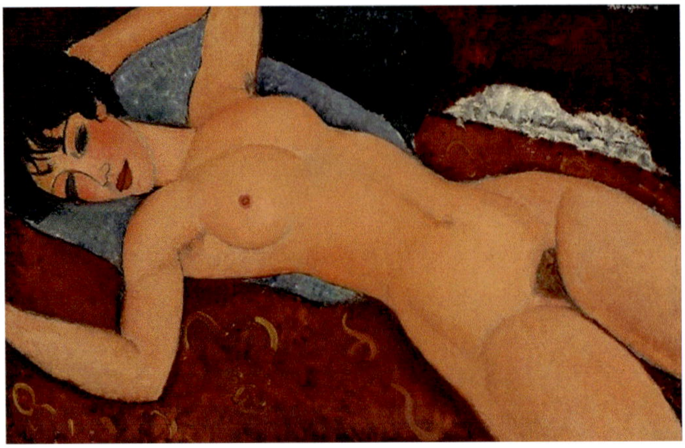

**Fig. 1.4** *Nu couche* (**Modigliani**). By Amedeo Modigliani—Christie' sale, New York, 9 November 2015, Public Domain {{PD-art-US}}, https://commons.wikimedia.org/w/index.php?curid=44456944. (1917; Long Museum, Shanghai, China)

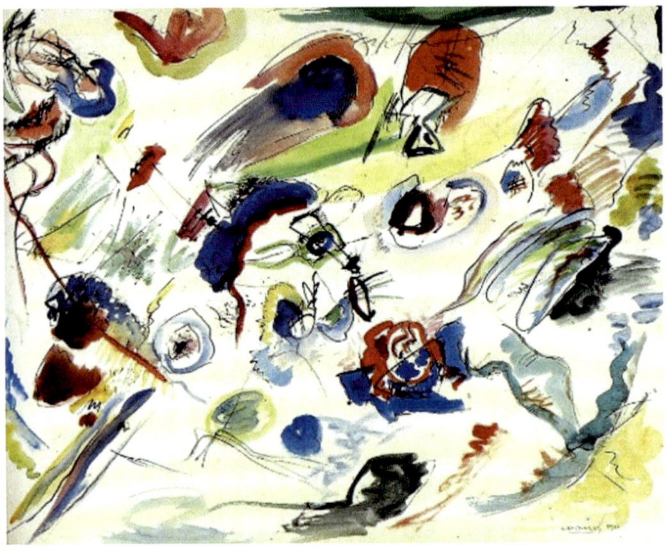

**Fig. 1.5** *Untitled* (**Kandinsky**) https://web.archive.org/web/20170108052708/ https://arthistoryproject.com/artists/wassily-kandinsky/untitled-first-abstract-watercolor/, Public Domain {{PD-art-US}}, https://commons.wikimedia.org/w/index.php?curid=89535292. 1913; Musée national d'art moderne, Centre Georges Pompido

In the course of using your familiar tools to characterize great works of art, you'll pick up a smattering of appreciation and recognition of many of the great artists: Michaelangelo, daVinci, Raphael, Bosch, Modigliani, van Gogh. And who knows how far your horizons may be expanded? Take a look at Figs. 1.5, 1.6, 1.7 and 1.8.

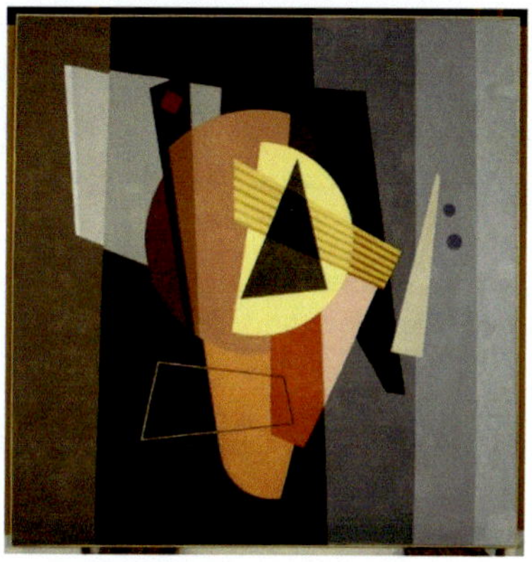

**Fig. 1.6** *Untitled* (**Kelpe**) online collection of Brooklyn museum; Photo: Brooklyn museum, L1990.1.3_SL1.jpg, public domain, https://commons.wikimedia.org/w/index.php?curid=14606715. 1938, Brooklyn museum, New York

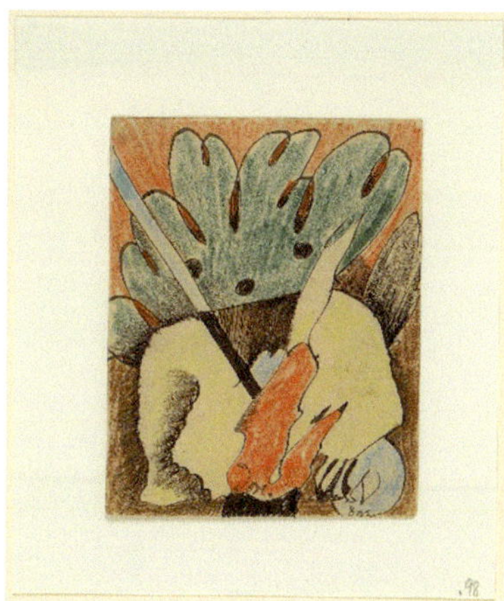

**Fig. 1.7** *Untitled* (**Dove**) by Arthur Dove—This file was donated to wikimedia commons as part of a project by the metropolitan museum of art. See the image and data resources open access policy, CC0, https://commons.wikimedia.org/w/index.php?curid=60912487

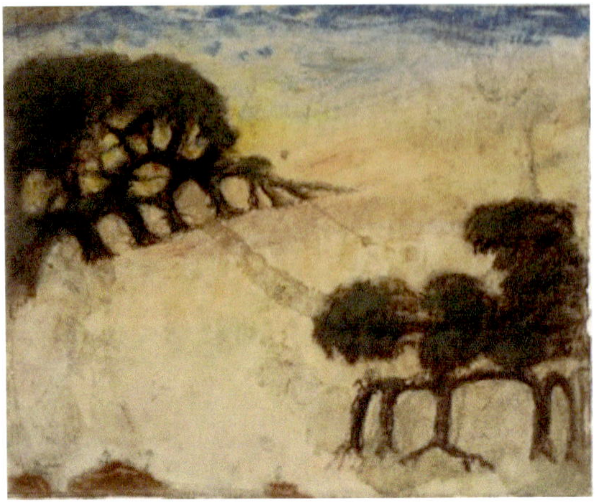

**Fig. 1.8** Untitled (**Hill**). By Carl Fredrik Hill—Malmö art museum, public domain {{PD-art-US}}, https://commons.wikimedia.org/w/index.php?curid=7669917. (1883)

See? One short chapter, and you've already memorized the names of four famous paintings!

# 2 Sculpture

## 2.1 Michelangelo's David

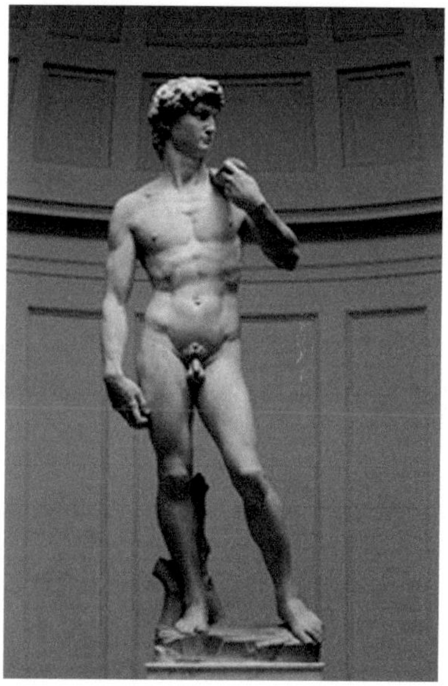

**Fig. 2.1** *David* (**Michelangelo**). By Jörg Bittner Unna—own work, CC BY 3.0, https://commons.wikimedia.org/w/index.php?curid=56633968. c. 1504; Galleria dell'Accademia, Florence

(Fig. 2.1)[1] A few years before the turn of the sixteenth century, stonecutters working in the hills of Tuscany were able to extract a huge block—17 feet long—of Carrara marble,

[1] Creative Commons License Links for all images are listed in the Preface.



A. D. Snider, *From STEM to STEAM*, Synthesis Lectures on Engineering, Science, and Technology, https://doi.org/10.1007/978-3-031-57316-3_2

intact, from the quarries. They carefully transported their prize to boats on the Arno River[2] and floated it down to Florence, where most of the famous sculptors of the Renaissance worked; they were sure they would get a large fee for such a massive block. And they were right. The City of Florence bought the marble and decided to have it carved into a statue of David.

Several master Florentine carvers were offered the commission for the project. Two (Agostino di Duccio and Antonio Rossellino) even began to cut the stone. But they all gave up. None of them had the confidence that they could execute a sculpture as delicate as the human body, on such a grand scale, without its collapsing under its own weight. The city decided to trim the block down to workable dimensions and re-commission the work accordingly.

But one young, unproven artist named Michelangelo Buonarroti (1475–1564) convinced the officials to allow him to try to carve the original piece. (In fact, Michelangelo had received some celebrity for his recent work, the Pieta—to be described.) It would be a transgression, he argued, to squander the opportunity to exploit this rare marble behemoth. The officials relented and gave the block to Michelangelo in 1501. In 1504, at the age of 29, Michelangelo unveiled his masterpiece—David—the best known sculpture in the world today. Originally intended as one of a series of statues of prophets to be positioned along the roofline of the Florentine cathedral, the stunning sculpture was placed instead in a public square outside the Palazzo della Signoria, the seat of government. It soon became the symbol of Florence, celebrating the defense of civil liberties embraced by the Republic.

We're geeks, so let's determine how fragile the sculpture really is. We'll estimate the stress at its critical points and compare it to the maximum stress that marble can bear.

Where would we find the highest points of stress in a statue of a (nude) man? Stress is proportional to force, of course; and the force in question is the weight of the sculpture. Since a man's legs have to carry almost all of his weight, we focus our search on the lower extremities.

However, the stress is less if the force is distributed over a wider area. In fact, stress is the ratio of the force to the cross-sectional area over which the force is distributed (pounds per square inch). Now a man's ankles have the smallest cross section of any part of the lower extremities, and they have to support practically the entire weight (everything but the feet), so the stress on the ankles is the crucial parameter that Michelangelo would have to confront for his huge undertaking. Right?

Wrong. Look more carefully (Fig. 2.2).

Do you see the "art-engineering" that Michelangelo did? He reinforced David's right ankle with a small tree trunk—increasing the area, redistributing the weight, and thereby decreasing the stress![3]

---

[2] An interesting description of the project can be found at http://stories-in-stone.blogspot.com/2008_11_01_archive.html.

[3] Bluffton College's site http://www.bluffton.edu/~sullivanm/micheldavid/david.html contains many excellent photographs of David. Pay particular attention to http://www.bluffton.edu/~sullivanm/micheldavid/legs2.jpg.

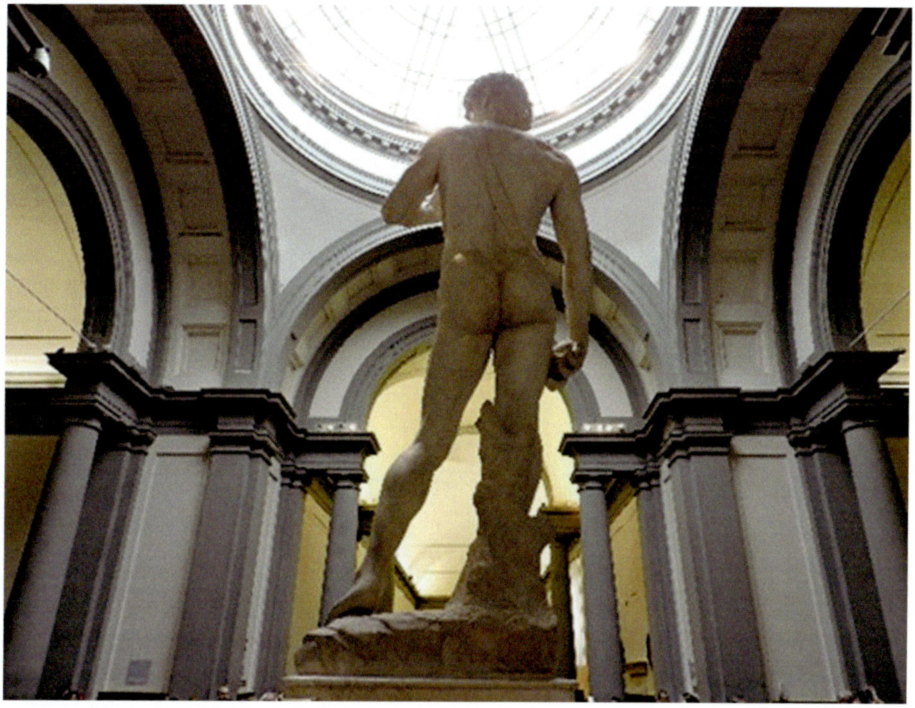

**Fig. 2.2** ***David*'s critical stress points** (Michelangelo). By Dimitris Kamaras from Athens, Greece—Michelangelo's David, Galleria dell'Accademia, Florence, CC BY 2.0, https://commons.wikimedia.org/w/index.php?curid=118173034. c. 1504; Galleria dell'Accademia, Florence

*Now*, where are the points of highest stress? The next smallest cross section in the legs is at the knees, which still have to support a great deal of the body's weight. So let's estimate the stress on David's knees, and compare it to the maximum compressive stress that marble can bear.

## 2.2 Estimating the Stress on David's Knees

To estimate the stress—force per unit area—on David's knees we need to estimate the weight of the statue from the knees upward, and the cross sectional area of the knees. Let's plan how to do this.

David is, of course, *geometrically similar* to a flesh-and-blood man; that is, all of his dimensions are proportional to those of a youth of comparable build. If you took, say, a Chippendales dancer and rescaled his dimensions—his height, his width, and his front-to-back depth—you would have a reasonable facsimile of David. And the scaling factors for height, width, and depth would be the same. This gives us a *Scaling Law*:

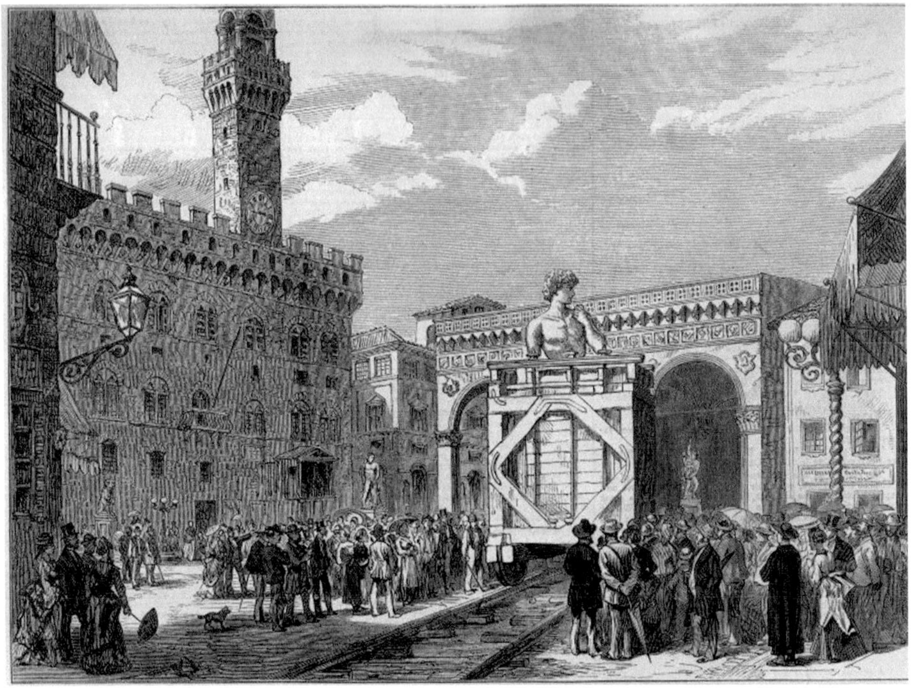

**Fig. 2.3** Moving Michelangelo's statue of David from Piazza della Signoria to Galleria. By unknown—illustrated London News (1873), Public Domain {{PD-US}}, https://commons.wikimedia.org/w/index.php?curid=137600457

If the ratio of David's height to a man's height is $r$, then

(1) the ratio of David's volumes (height-by-width-by-depth) to the man's volumes is $r^3$;
(2) the ratio of David's cross-section areas (width-by-depth) to those of the man is $r^2$.

David's stomach holds $r^3$ times as much as the man's stomach, David's footprint covers $r^2$ times the man's footprint, etc. This is obviously true for rectangular volumes ($h \times w \times d$) and areas ($w \times d$), spherical volumes ($\frac{4}{3}\pi R^3$) and circular areas ($\pi R^2$); it is proved in general using calculus.

One can estimate the height ratio for David by looking at a drawing from the Illustrated London News (1873), which depicts the statue being moved to its current location. (Fig. 2.3. We'll give the story behind this picture in the next section.)

It appears that David is about as tall as three men. In fact David is 517 cm tall, or 2.83 times taller than a 6-ft man.[4] So we'll take $r = 2.8$.

[4] http://totallyhistory.com/david-by-michelangelo/; https://graphics.stanford.edu/papers/digmich_falletti/; https://accademia.stanford.edu/mich/.

David's weight is distributed over the area of David's knees, which is $r^2 = 2.8^2$ times the area of a man's knees. If you (guys) take a measuring tape and measure the circumference of your knee at its narrowest point, you'll get about 13 in. The circumference around David's knees is then $2.8 \times 13 = 36.4$ in. That gives a radius of $R = 36.4/(2\pi) = 5.8$ in., and an area of $\pi \times 5.8^2 = 106$ in.$^2$. For a biped like David, the total knee cross-section area is about 212 in.$^2$.

(In the interest of modesty: in 1857 a fig leaf was designed for the full-sized replica of David residing at the Victoria and Albert Museum in London.[5])

The force that is distributed over this area is David's weight above the knees. David's calves and feet probably account for about one-eighth of his body weight. Therefore the knees support 7/8 of the body weight. This weight equals the density of marble (170 lbs/ft$^3$) [6] times the volume. David's volume is hard to estimate, but we can finesse it by noting that it will be $2.8^3$ times the volume of a 6 ft man, and the *man*'s volume equals *his* weight divided by *his* density.

Since men just barely float in water, their density is roughly the same as that of water (62.4 lb/ft$^3$).[7] A 6-ft man with the build of a Chippendales dancer weighs about 190 pounds. Therefore the volume of the Chippendale is roughly

$$190\ \text{lbs}/(62.4\ \text{lb}/\text{ft}^3) = 3.0\ \text{ft}^3,$$

and David's volume is

$$2.8^3 \times 3.0\ \text{ft}^3 = 66\ \text{ft}^3.$$

David's weight is thus

$$\text{density times volume} = 170\,\text{lbs}/\text{ft}^3 \times 66\,\text{ft}^3 = 11{,}220\,\text{lb},$$

and the force on his knees is 7/8 of this. Therefore the stress is roughly

$$\text{weight}/\text{area} = (7/8) \times 11{,}220\ \text{lbs}/212\ \text{in.}^2 = 46.3\ \text{lb}/\text{in.}^2.$$

References[8] cite a compressive strength for marble around 10,000–20,000 lb/in.$^2$. David is not going to collapse at the knees!

---

[5] http://www.vam.ac.uk/content/articles/d/davids-fig-leaf/. (If you have to know: the leaf was 19 inches long.)

[6] http://www.aqua-calc.com/page/density-table/substance/marble-coma-and-blank-solid.

[7] http://www.aqua-calc.com/page/density-table/substance/water.

[8] http://www.alabamamarblequarry.com/index.php/component/content/article/10-press-release/8-test-certificate; https://pubs.naturalstoneinstitute.org/pub/2c4ec57c-aef5-8a85-16e0-106c5cada13c

## 2.3 Transporting Michelangelo's Sculptures

The transportation of David was a thorny problem for the Florentines. They had to move him very carefully, on a large cart, from Michelangelo's workshop to the main square, the Piazza della Signoria. During the transport a piece of his forearm was knocked loose. The story goes that the arm was recovered by a boy named Giorgio Vasari, who became a very important figure in the history of Florence. We'll come back to Vasari.

David is far too fine a piece to be exposed to the elements, so in 1873 he was moved to the Accademia di Belle Arti (Fig. 2.3), where he resides to this day. Special rails were installed along Florence's streets for this move, and it proceeded without incident.[9]

But there's more to the story. In 1964 the World's Fair was held in Queens, New York, and Pope John XXIII decided to allow one of Michelangelo's sculptures to be shipped overseas for the Vatican Pavilion.[10]

They didn't take a chance with David, but instead chose the *Pieta*, Michelangelo's initial acclaimed masterpiece that resides in the Vatican in Rome (Fig. 2.4).

How do you cushion such a work against the possibility, e.g., of storms at sea?

Predictably it was packaged inside two extremely sturdy concentric cases, each cushioned with millions of tiny polystyrene beads, and then wrapped in asbestos.[11] Pieta was insured for $6,000,000 while undertaking this voyage. (Later the Vatican Museum decided never again to risk lending out such a piece.)

An interesting proposal was made to *suspend* the Pieta from the eight corners of a steel-beam cube by cables, so that there would be no possibility of its colliding with any hard surface. This strategy was not carried out, although a news photo of Pieta's passage on the cruise ship Christoforo Columbo does indicate someone was thinking along these lines.[12] (In 2006, a freshman attending your author's class at the University of South Florida won the Engineering College's Egg Drop contest by cushioning his egg in exactly this manner!)

This suggests another calculation (Fig. 2.5): would David break if he were suspended upside down?

We don't have to do a lot of work; the knees will still be the weak point, and they will still carry 7/8 of the body weight. But now the sculpture is in *tension*, rather than *compression*. As before the knee stress equals about 46.3 lb/in.$^2$; and this must be compared to marble's *tensile* stress limit, which is only around 10% of the compressibility limit

[9] This video will amuse you: https://www.youtube.com/watch?v=yvvES47OdmY.

[10] https://en.wikipedia.org/wiki/1964_New_York_World%27s_Fair#/media/File:New_York_World%27s_Fair_August_1964.jpeg.

[11] An account of the preparations can be found at http://nywf64.com/vatican08.html.

[12] Ibid.

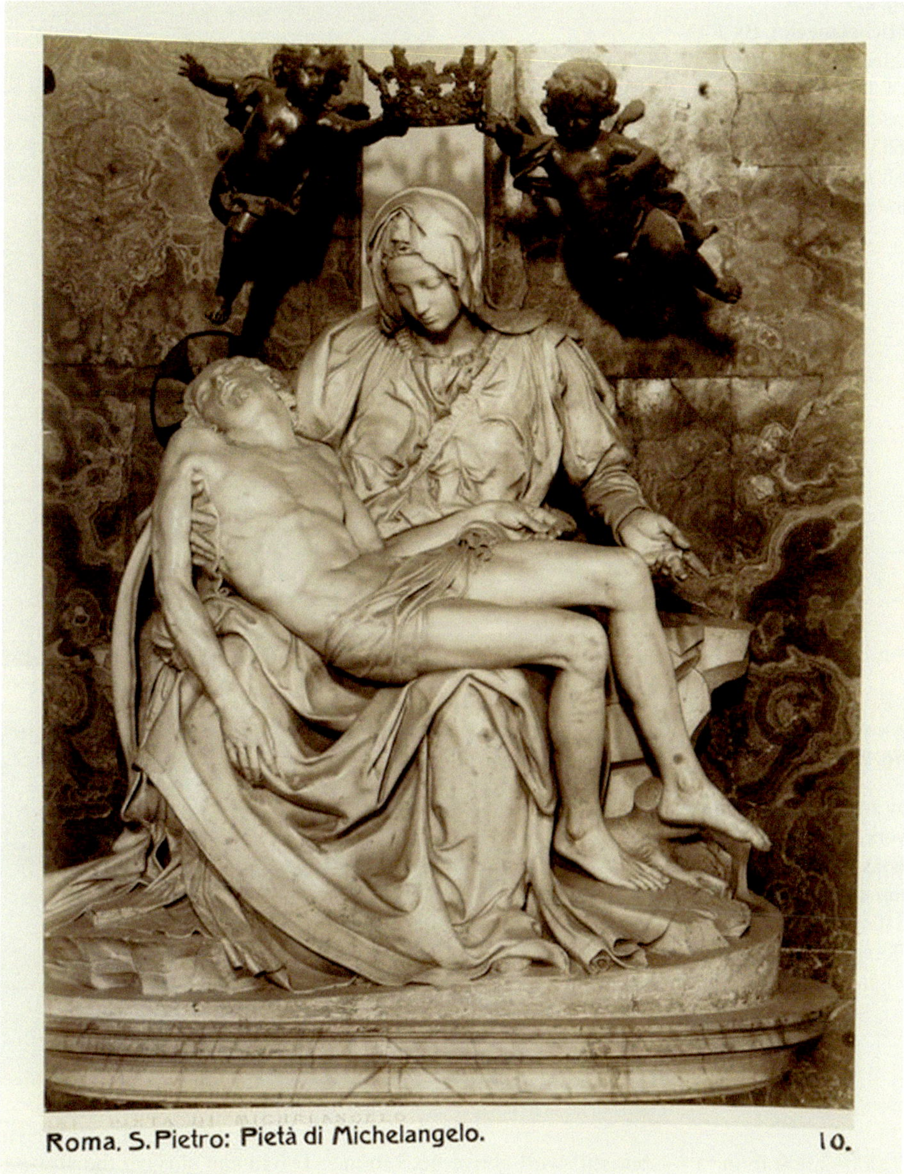


**Fig. 2.4** *Pieta* (**Michelangelo**) by unknown author—LSH 107520 (hm_dig18719), Public Domain, {{PD-1996}}; https://commons.wikimedia.org/w/index.php?curid=53704728. Photographed by Jenny Bergensten. 1499, Saint Peter's Basilica, Vatican City

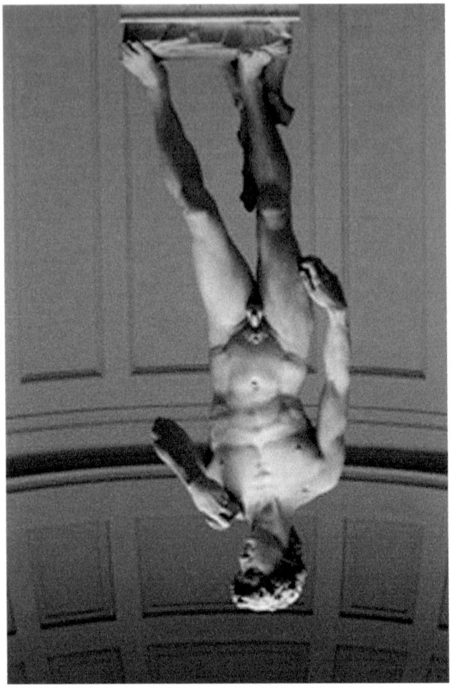

**Fig. 2.5** David (**Michelangelo**). By Jörg Bittner Unna—own work, CC BY 3.0, https://commons.wikimedia.org/w/index.php?curid=56633968. c. 1504; Galleria dell'Accademia, Florence. Embellished from original

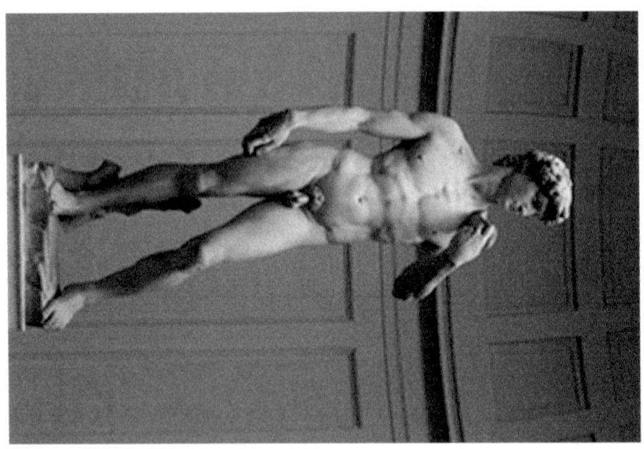

**Fig. 2.6** David (**Michelangelo**). By Jörg Bittner Unna—own work, CC BY 3.0, https://commons.wikimedia.org/w/index.php?curid=56633968. c. 1504; Galleria dell'Accademia, Florence. Embellished from original

(1,000–2,000 lb/in.$^2$)[13]—but still well above 46.3 lb/in.$^2$. David can survive the stress (if not the indignity) of being dangled in this manner. But it's safer to transport him upright.

Just to flex our engineering muscles let's explore the limits of David's indestructibility by estimating the stresses if he were mounted sideways (Fig. 2.6).

[13] http://www.globalguideline.com/interview_questions/Answer.php?a=What_is_the_tensile_strength_of_marble; http://www.alabamamarblequarry.com/index.php/component/content/article/10-press-release/8-test-certificate.

David's left knee will be in compression and his right knee in tension; we have to make a torque computation. The exercise below guides you through the math, and shows that the stresses are quite high. In fact the tensile stress on the right knee is roughly 10–20% of the maximum allowable! Not surprisingly, this is a terrible way to transport David—one nasty road bump, and you'd have a lot of reverse engineering to do.

**Exercise** You can estimate the compression on David's left knee by balancing the torques—lever arm times force—about his right knee. The key equation is $d_1 F_1 = d_2 F_2$, where

$d_1$ = distance from David's right knee to his center of mass,
$F_1$ = weight of David's body above the knees,
$d_2$ = distance between David's knees, and
$F_2$ = compressive force on David's left leg.

$F_2$ is the unknown. $d_1$ is probably about 40% of David's height, which we estimated to be about $r = 2.8$ times 6 feet. We have estimated $F_1$ to be (7/8) $\times$ 11,220 pounds.

To estimate $d_2$ we will use the scale factor $r$ times the corresponding distance for a 6-ft man. Have your guy assume a stance like David's and measure the distance between his knees.[14]

Plug the data into the equation and solve for $F_2$. This is the compressive force on David's left leg, which we may assume is roughly the same as the tensile force on the right leg. Divide by 212 $\text{in.}^2$, which we estimated to be the cross section area of David's knees, to get the stress.

You will find that your answer is well below marble's compressive stress limit (10,000–20,000 $\text{lb/in.}^2$), but uncomfortably close to the tensile stress limit (1,000–2,000 $\text{lb/in.}^2$).

Several structural engineers have studied these stresses more rigorously using computer simulation and laser photography.[15] Figure 2.7 displays some results calculated by the Spatial Automation Laboratory at the University of Wisconsin-Madison (V. Shapiro et al.). The "von Mises" *tensile* stresses are listed in Pascals (1 Pa = 0.000145 $\text{lb/in.}^2$); thus *compressive* stresses are depicted as negative. Observe that our crude estimate of the compressive stress on the knees, 46.3 $\text{lb/in.}^2 \approx 2.87 \times 10^5$ Pa, is in rough agreement with the calculated value for the right knee. The figure also shows that Michelangelo was wise to reinforce the right ankle with the tree trunk; the stress is nearly $3 \times 10^6$ Pa, and is tensile! This is due to the fact that David is actually leaning forward. Indeed, the Stanford graphics revealed hairline cracks in the ankles.

[14] (He can keep his clothes on.)

[15] http://graphics.stanford.edu/projects/mich/other-body-parts/other-body-parts.html; https://www.sciencedirect.com/science/article/pii/S1296207406000768; https://www.sciencedirect.com/science/article/pii/S0168874X11000448.

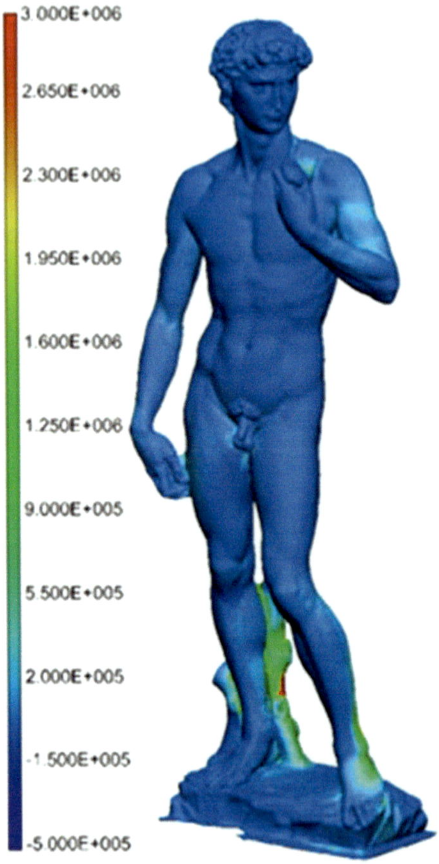


**Fig. 2.7** Stress diagram for Michelangelo's *David*. https://www.nsf.gov/news/mmg/media/images/statue_david3_h1.jpg; *Credit* Members of the Spatial Automation Laboratory, University of Wisconsin-Madison. Reproduced by permission of V. Shapiro

## 2.4 Other Davids

What was the best known sculpture in the world *before* Michelangelo created David? A likely answer is an earlier (1430) David by Donatello (Donato di Niccolò di Betto Bardi, 1386–1466). It also resides in Florence (Bargello Museum), but it differs in many ways from Michelangelo's version; it was executed in bronze, not marble, and it is only a little over five feet tall. (No structural issues here!)

Both Davids share a certain degree of effeminacy. It is quite pronounced in the Donatello David. Look first at the *bust* of the statue (Fig. 2.8).

The sweet face and the laurel on the helmet often convey the impression of a young girl to viewers; I was a little startled when I lowered my gaze (Fig. 2.9).

But make no mistake; photographs cannot do this sculpture justice. In the flesh (well, bronze), it is beautiful.

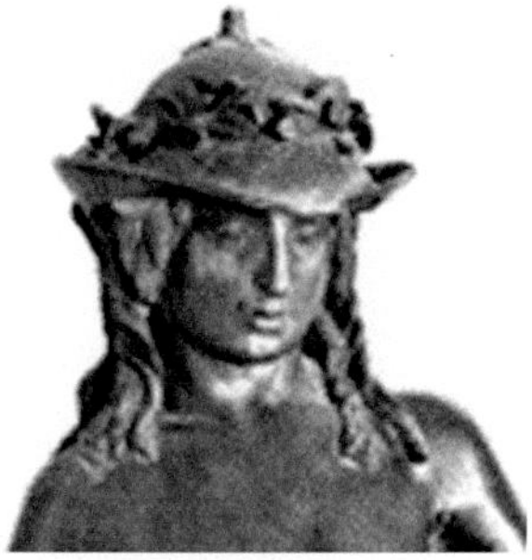

**Fig. 2.8** *David* (**Donatello**). By ketrin1407—Donatello (c.1386–1466)—David (date disputed) front, CC BY 2.0, https://commons.wikimedia.org/w/index.php?curid=79313474. 1440; Bargello, Florence. Cropped from original

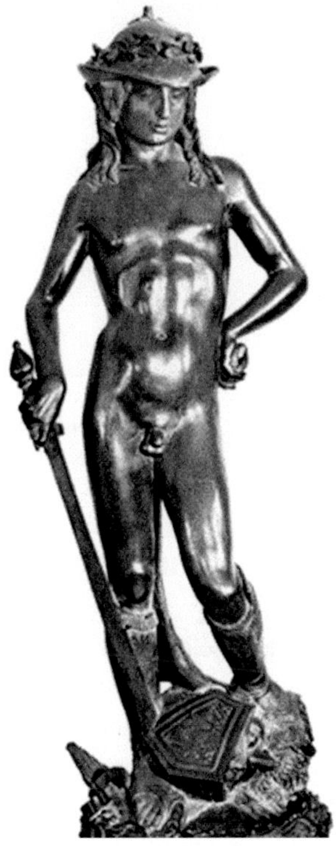

**Fig. 2.9** *David* (**Donatello**). By ketrin1407—Donatello (c.1386–1466)—David (date disputed) front, CC BY 2.0, https://commons.wikimedia.org/w/index.php?curid=79313474. 1440; Bargello, Florence

A totally different image of David was created by Gian Lorenzo Bernini (1598–1680) in 1623. Look at this David's face (Fig. 2.10). No shrinking violet he, Bernini's David is set to "kick ass and take names."[16]

[16] J. O. Killens coined this phrase in *Youngblood*, U. Georgia Press, Athens, GA; (2000), ISBN: 9-780-8203-2201-8, p. 426.

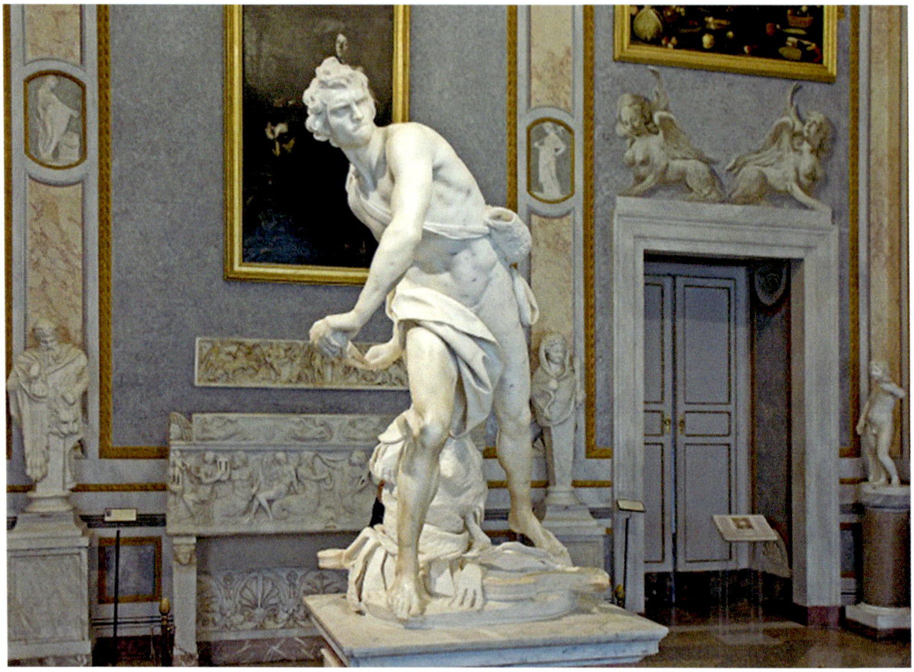

**Fig. 2.10** *David* (**Bernini**). By Jk1677—own work, CC0, https://commons.wikimedia.org/w/index.php?curid=45633866. 1624; Borghese, Rome

## 2.5 Other Michelangelos

Michelangelo's best known sculptures are David, the Pieta, and Moses (Fig. 2.11). You have seen the first two. Your author would argue that *Moses* is the greatest. (I swear I once thought the sculpture moved while I was viewing it.) Enlarge the image in Fig. 2.12. Look at the furrowed brow, the muscles and sinews, even the toenails! Moses is part of the tomb of Pope Julius, an ambitious project that Michelangelo never completed. It resides in the church of St. Peter in Chains in Rome. (And it shares a wall with the Engineering College of La Sapienza University!)

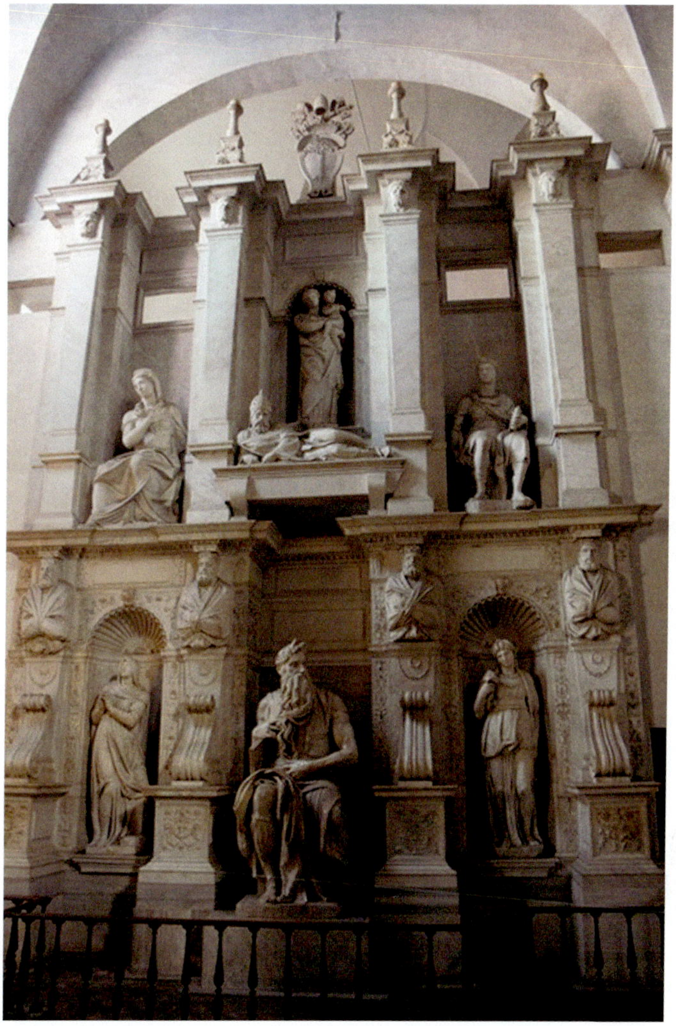

**Fig. 2.11** *Moses* (**Michelangelo**) by Szilas—own work photo by Szilas, Public Domain, https://commons.wikimedia.org/w/index.php?curid=36741687. 1505–1545; Basilica of St. Peter in Chains, Rome

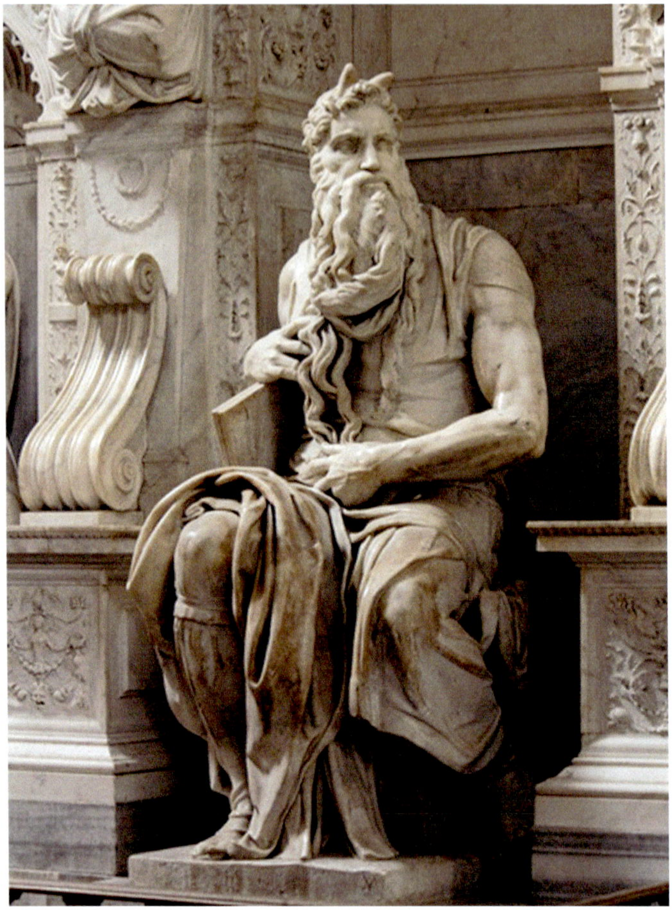

**Fig. 2.12** *Moses* (**Michelangelo**) by Jörg Bittner Unna—own work, CC BY 3.0, https://commons.wikimedia.org/w/index.php?curid=35823159 1505–1545; Basilica of St. Peter in Chains, Rome

The protuberances on Moses' forehead are intriguing. Moses was said to have shared a direct "line of contact" with God, and these bumps are truncations of *rays to heaven.* They are depicted in many other Moses representations (Figs. 2.13 and 2.14).

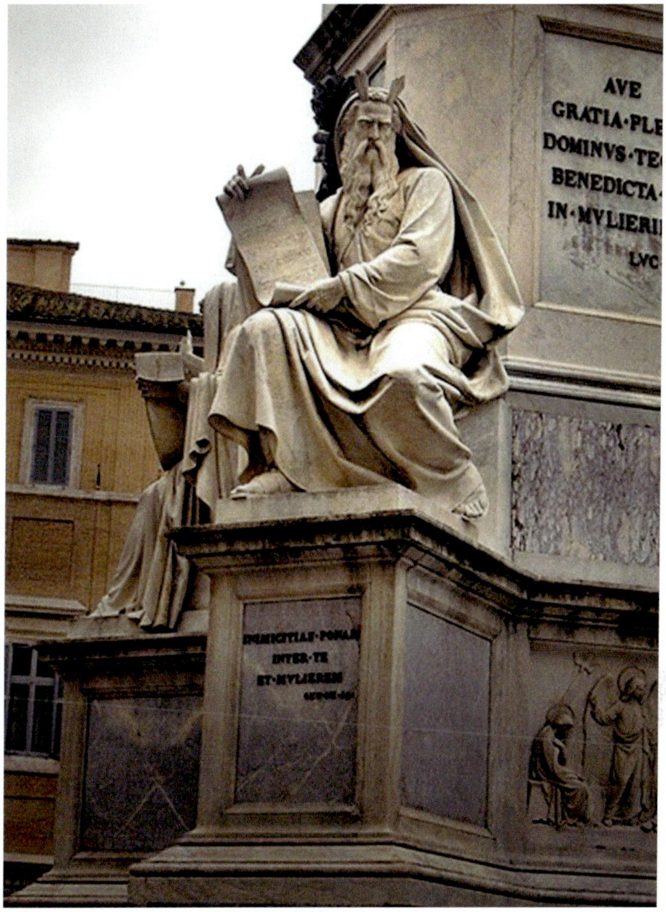


**Fig. 2.13** *Immaculate conception column* detail (**Poletti**). By Georges Jansoone (JoJan)—own work, CC BY 3.0 (https://creativecommons.org/licenses/by/3.0/deed.en), https://commons.wikimedia.org/w/index.php?curid=10233693. 1857; Piazza Mignalli, Rome

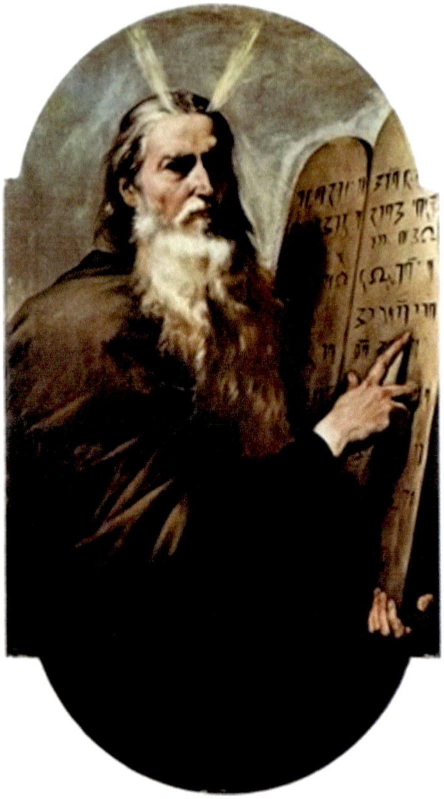

**Fig. 2.14** *Ten Commandments* (**Ribera**). By Jusepe de Ribera—The Yorck Project (2002) 10.000 Meisterwerke der Malerei (DVD-ROM), distributed by DIRECTMEDIA Publishing GmbH. ISBN: 3936122202. Public Domain, https://commons.wikimedia.org/w/index.php?curid=2411717. 1638, Museo Nazionale di San Martino, Naples

The deliberate distortion that Moses had "horns" appeared in early antisemitic art (Fig. 2.15). (See also https://medium.com/lessons-from-history/the-intriguing-interpretation-of-horns-in-michelangelos-sculpture-moses-a1044179408b.)

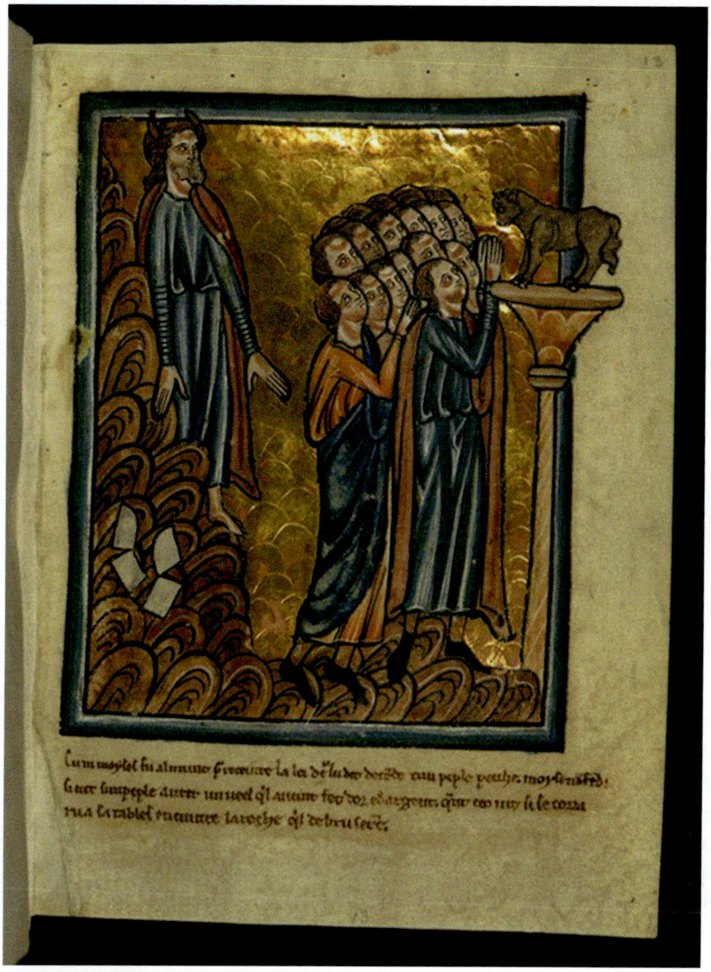

**Fig. 2.15** Illustration of Moses with horns from a 13th-century illuminated manuscript (**William de Brailes**). https://commons.wikimedia.org/wiki/File:William_de_Brailes_-_The_Israelites_Worship_the_Golden_Calf_and_Moses_Breaks_the_Tablets_(Exodus_32_-1-19)_-_Walters_W10613R_-_Full_Page.jpg. c.1250; Walters Art Museum. Baltimore. Public Domain, via Wikimedia Commons. {{PD-1996}}

Other notable sculptures by Michelangelo (and his pupils)[17] are the *Four Prisoners* at Florence's Accademia di Belle Arti, the *Rebellious Slave* and the *Dying Slave* in the Louvre, the *Deposition* or *Florence Pieta* at the Duomo in Florence, and the Capitani Tombs in the Medici Chapels in Florence (Figs. 2.16, 2.17, 2.18, 2.19, 2.20, 2.21 and 2.22). The *Prisoners* are particularly interesting because, although they are unfinished, they convey an eerie sense that their subjects are emerging from the marble.

[17] For more of Michelangelo's sculptures consult https://entertainment.howstuffworks.com/arts/artwork/michelangelo-sculptures.htm#pt11.

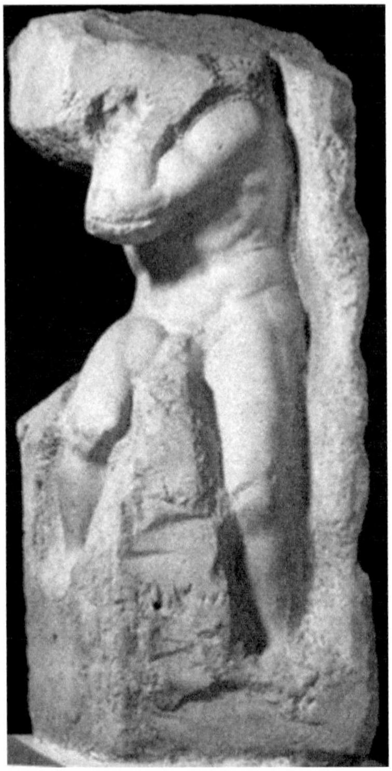

**Fig. 2.16** *Atlas Slave* (**Michelangelo**). Гнедич. "История искусств", Public Domain {{PD-Russia-expired}}, {{PD-US}}, https://commons.wikimedia.org/w/index.php?curid=10870794. 1523; Galleria dell'Accademia, Florence. Cropped from original

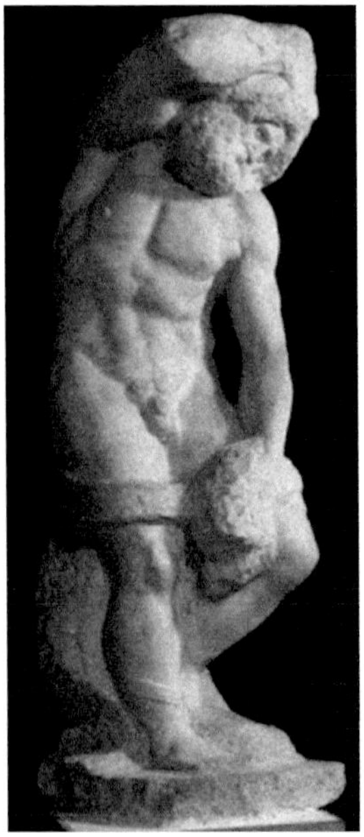

**Fig. 2.17** *Bearded Slave* (**Michelangelo**). By Anonymous—Гнедич. "История искусств", Public Domain, https://commons.wikimedia.org/w/index.php?curid=10870779. c. 1501; Galleria dell'Accademia, Florence. Public Domain {{PD-Russia-expired}}, {{PD-US}}). Cropped from original

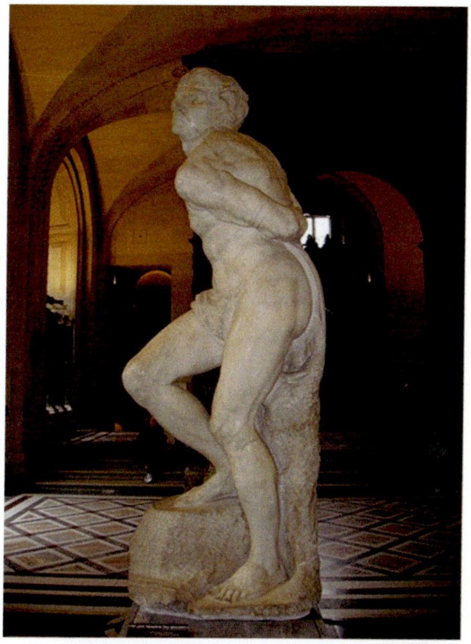

**Fig. 2.18** *Rebellious Slave* (**Michelangelo**) Public Domain, https://commons.wikimedia.org/w/index.php?curid=2746984. Photographed by Yair Haklai. c. 1513; Louvre; Paris

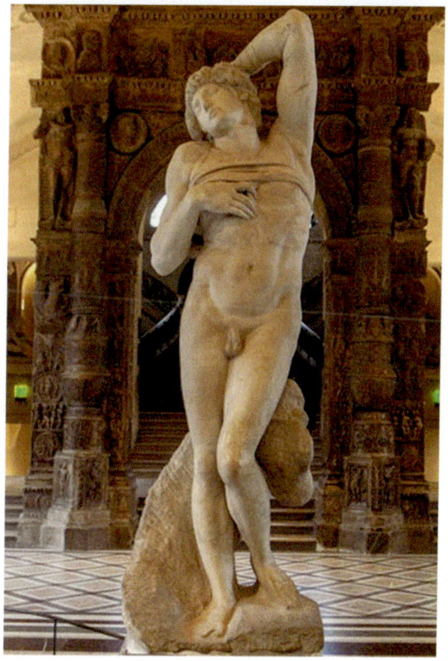

**Fig. 2.19** *Dying Slave* (**Michelangelo**) by Frank Kovalchek from Anchorage, Alaska, USA—The Slave, by Michelangelo, CC BY 2.0, https://commons.wikimedia.org/w/index.php?curid=24473420. c. 1515; Louvre, Paris

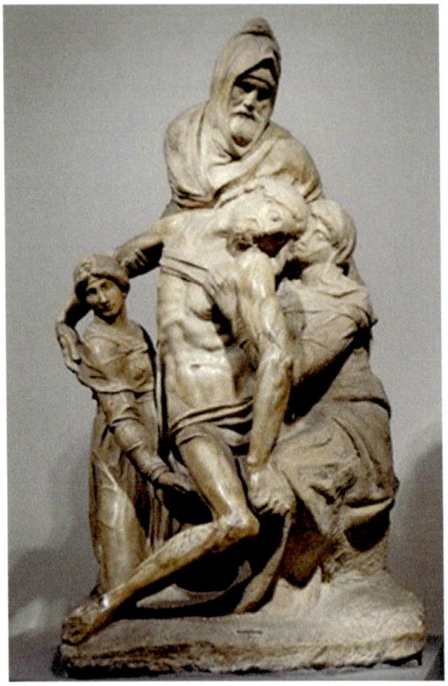

**Fig. 2.20** *Pieta* (**Michelangelo**) by © Marie-Lan Nguyen/Wikimedia Commons, CC BY 2.5 (https://creativecommons.org/licenses/by/2.5/deed.en). c. 1515; Museo dell'Opera del Duomo, Florence

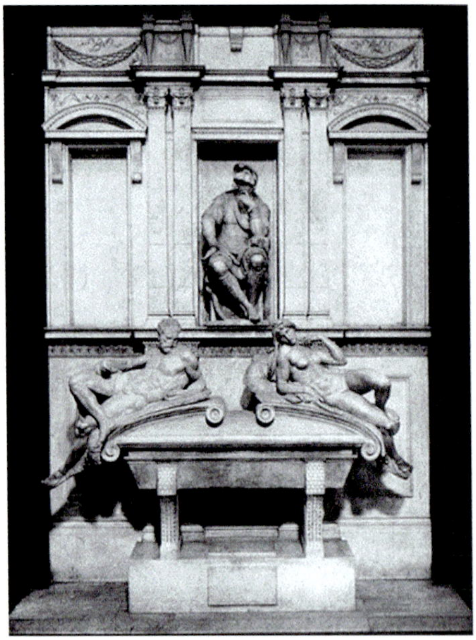

**Fig. 2.21** *Lorenzo de Medici* (**Michelangelo**) by Michelangelo—https://archive.org/details/lifeofmichaelang00roll (from JP2 files), by Romain Rolland, translated by Frederic Lees, 1912. Public Domain {{PD-US-expired}} https://commons.wikimedia.org/w/index.php?curid=15406004. c. 1533; Tomb of Lorenzo de Medici, Florence

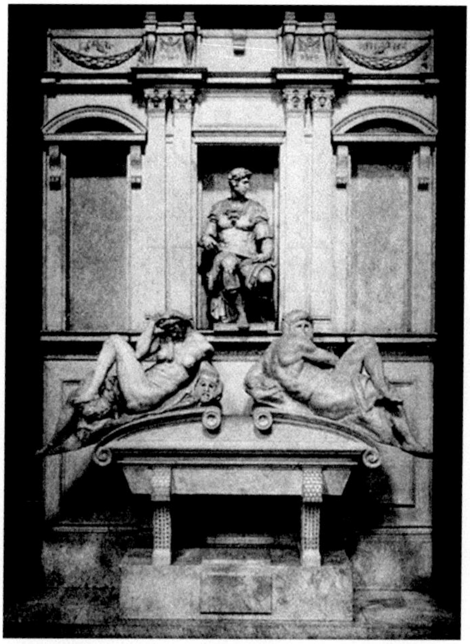

**Fig. 2.22** *Giuliano di Lorenzo de Medici* (**Michelangelo**) by Michelangelo—https://archive.org/details/lifeofmichaelang00roll (from JP2 files), by Romain Rolland, translated by Frederic Lees, 1912. Public Domain {{PD-US-expired}} https://commons.wikimedia.org/w/index.php?curid=15405900. c. 1533; Tomb of Giuliano de Medici, Florence

But Michelangelo's second-best known work is probably the Sistine Chapel in the Vatican (Figs. 2.23 and 2.24).[18] Though Michelangelo's first love was sculpture, in 1508 Pope Julius II insisted that he paint the chapel's ceiling (one of the reasons that he never finished Julius's tomb).

[18] See *Michelangelo and the Pope's Ceiling*, Ross King, Bloomsbury USA; Reprint edition 2014, ISBN-10: 1620408406 ISBN-13: 978-1620408407.

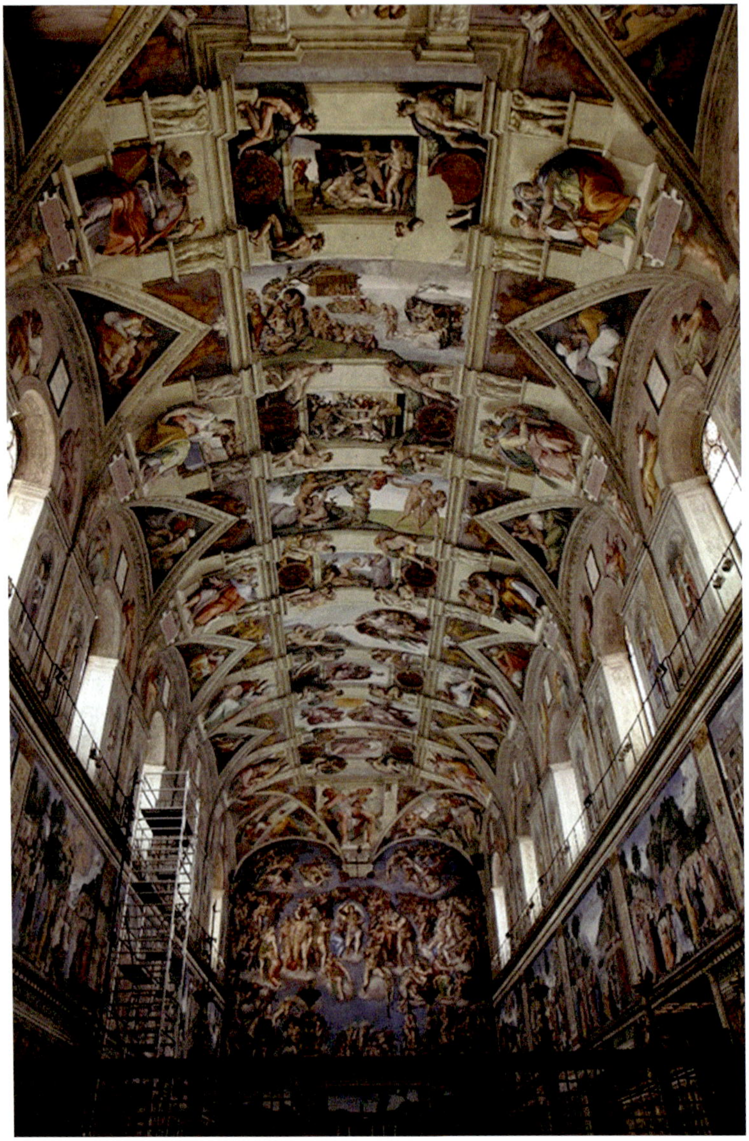

**Fig. 2.23** *Sistine Chapel* by Jörg Bittner Unna—own work, CC BY 3.0, https://commons.wikimedia.org/w/index.php?curid=46470792c. 1541; Sistine Chapel, Vatican City

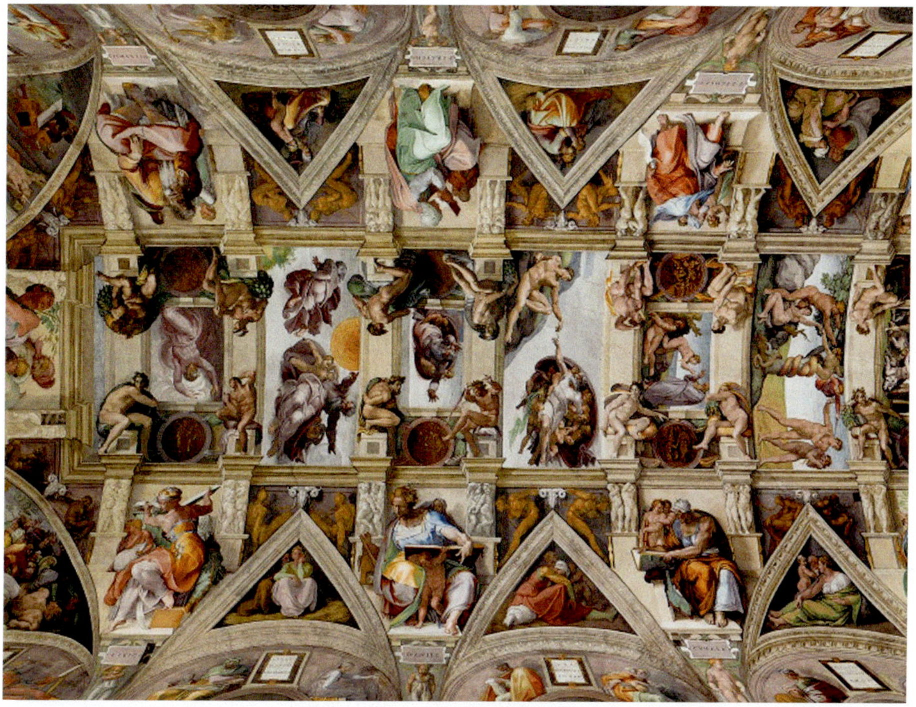

**Fig. 2.24** *Sistine Chapel Ceiling* (**Michelangelo**). By Aaron Logan, CC BY 2.5, https://commons.wikimedia.org/w/index.php?curid=13609. c. 1512, Sistine Chapel, Vatican City

Michelangelo built the scaffolding and painted the entire ceiling. You will recognize many of the well-known images in this work: the creation of Adam, the expulsion from the Garden of Eden, the creation of the sun and moon, and the prophet Isaiah (Figs. 2.25, 2.26, 2.27 and 2.28).

Michelangelo's Isaiah was the inspiration for **Norman Rockwell's** *Rosie the Riveter* poster during World War II.[19]

[19] (See http://www.mmdtkw.org/RenRom0418-SistIsiahRosie.jpg; https://en.wikipedia.org/wiki/Rosie_the_Riveter; https://www.loc.gov/wiseguide/feb04/rosie.html; or https://www.saturdayeveningpost.com/2013/07/rosie-the-riveter/.)

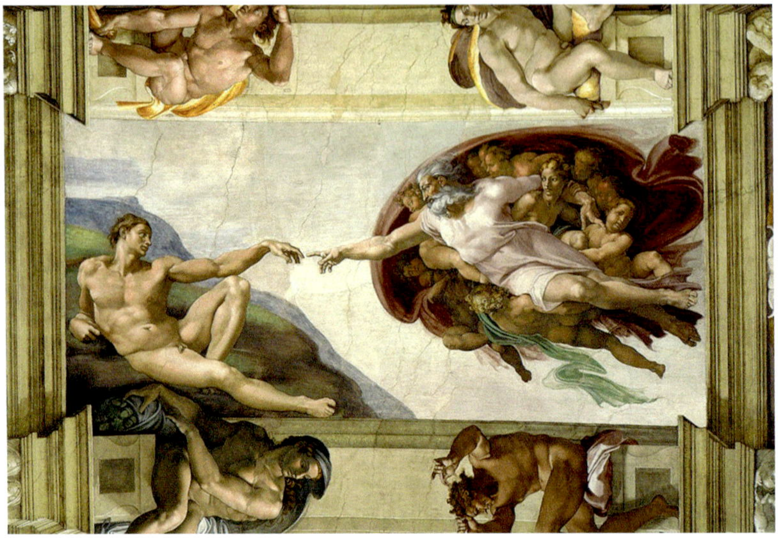

**Fig. 2.25** *Creation of Adam* (**Michelangelo**)—Public Domain (age), https://commons.wikimedia.org/w/index.php?curid=16022922. {{PD-US}} c. 1511, Sistine Chapel, Vatican City

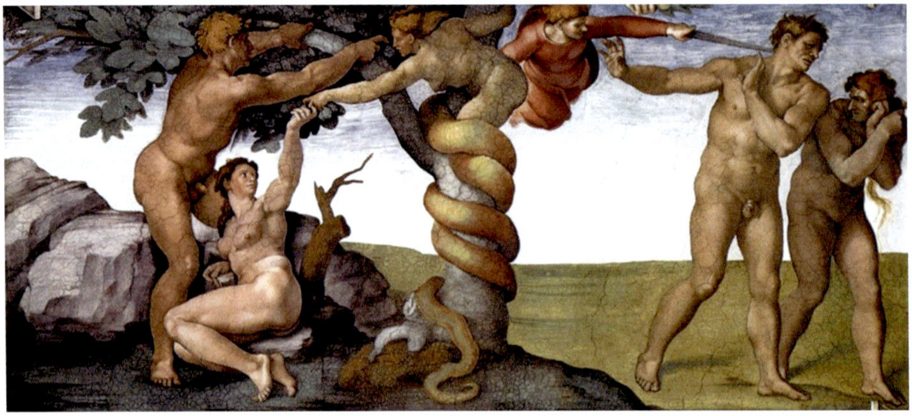

**Fig. 2.26** *Expulsion from the garden of Eden* (**Michelangelo**)—Web Gallery of Art[1], Public Domain (age), https://commons.wikimedia.org/w/index.php?curid=1551131. PDArt. {{PD-US}} c. 1511, Sistine Chapel, Vatican City

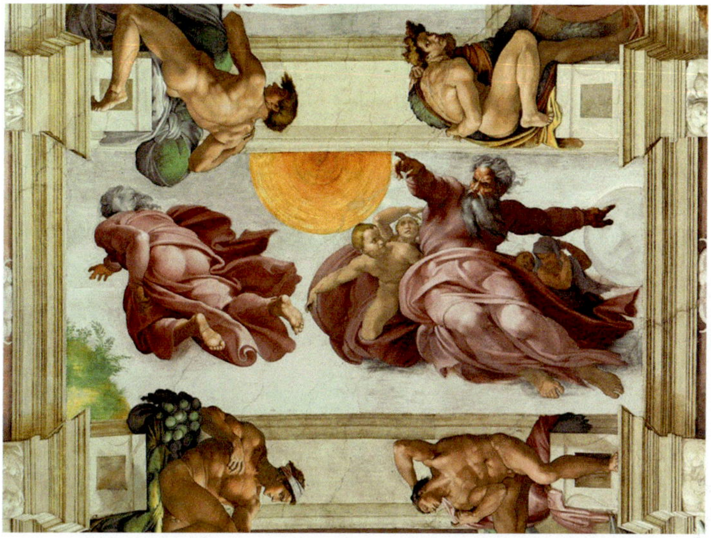

**Fig. 2.27** *Creation of the Sun and Moon* (**Michelangelo**)—Sistine Chapel, Public Domain (age), https://commons.wikimedia.org/w/index.php?curid=6084615 {{PD-US}} c. 1511, Sistine Chapel, Vatican City

**Fig. 2.28** The Prophet Isaiah; by Michelangelo—self-scanned, Public Domain (age), https://commons.wikimedia.org/w/index.php?curid=2776989. c. 1511, Sistine Chapel, Vatican City

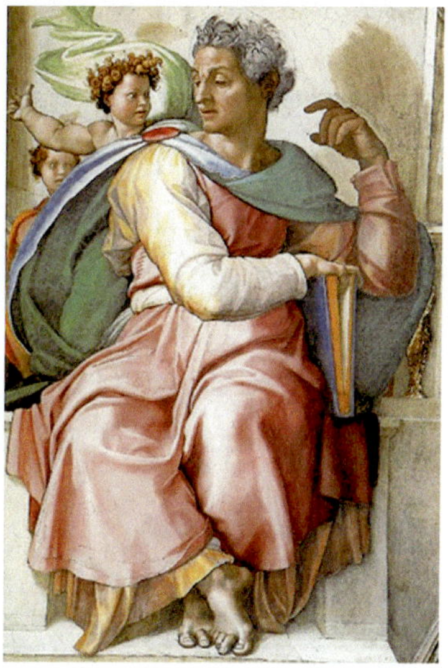

It took Michelangelo four years to complete the ceiling. He then returned to his other projects, including Julius' tomb, but in 1537 Pope Clement VII interrupted his plans by convincing him to paint the wall above the altar, with a depiction of The Last Judgment (Fig. 2.29). This astonishing painting required another four years.

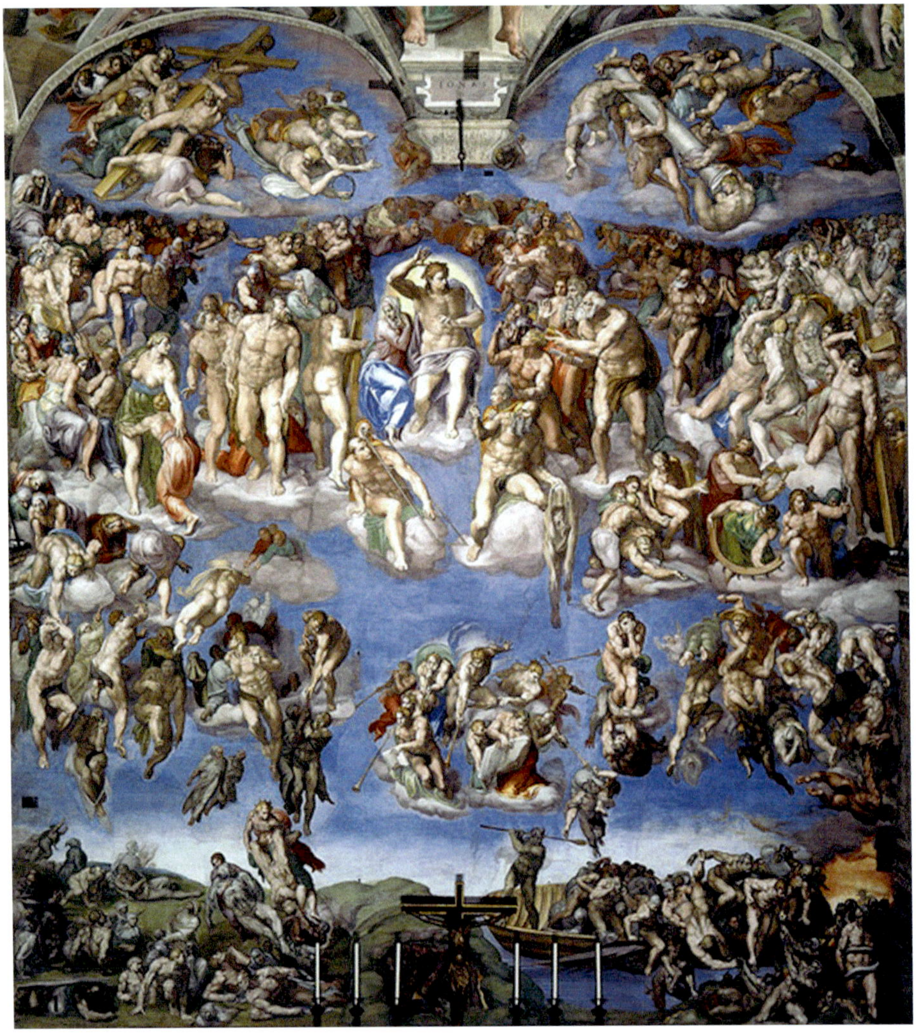

**Fig. 2.29** *Last Judgement* (**Michelangelo**)—Public Domain (age), https://commons.wikimedia.org/w/index.php?curid=16143987. c. 1541, Sistine Chapel, Vatican City

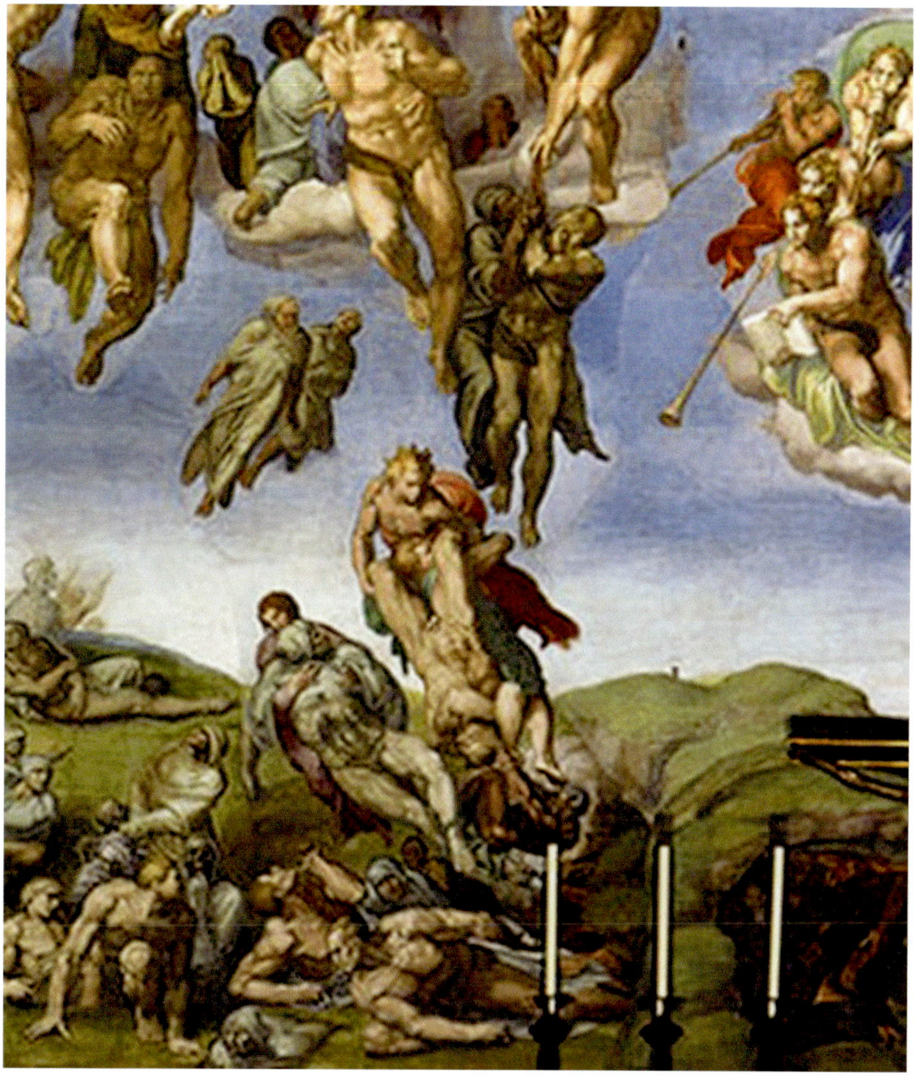

**Fig. 2.30** *Resurrection of the dead* (detail of Fig. 2.29) https://en.wikipedia.org/wiki/The_Last_Judgment_(Michelangelo)#/media/File:Michelangelo,_giudizio_universale,_dettagli_35.jpg; by see filename or category—www.wga.hu, Public Domain

Use your software or a magnifying glass to enlarge the details of Fig. 2.29 as we discuss them (Figs. 2.30, 2.31, 2.32 and 2.33). In the lower left corner, angels are trumpeting bodies out of their graves, to be judged. The righteous are whisked up to heaven, despite resistance by the demons of Hell. In the lower center, the condemned are ferried across the river Styx to Hades in Charon's boat. In the upper right center the flayed Saint,

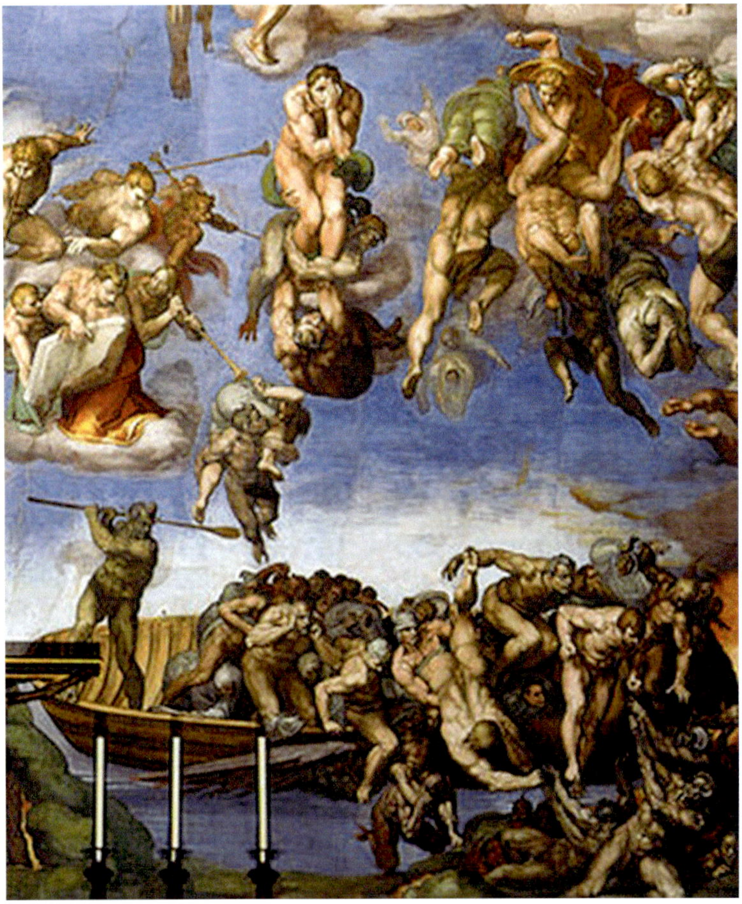

**Fig. 2.31** *Charon's Ferry* (detail of Fig. 2.29) https://en.wikipedia.org/wiki/The_Last_Judgment_(Michelangelo)#/media/File:Michelangelo,_giudizio_universale,_dettagli_51_inferno.jpg; by see filename or category—scan: De Vecchi, Cappella Sistina, 1999, Public Domain, https://commons.wikimedia.org/w/index.php?curid=11963135

Bartholomew, is shown holding his own skin, in which Michelangelo's face is depicted. (Do you see the resemblance of Bartholomew in Fig. 2.32 to the Florence Pieta sculpture in Fig. 2.20?) Just out of the frame of the upper right corner of Fig. 2.29, Jonah is shown with the "great fish" that swallowed him (Fig. 2.33).

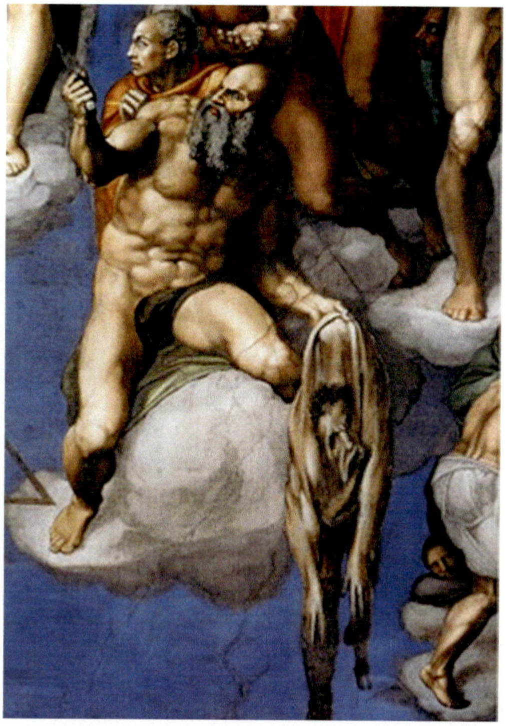

**Fig. 2.32** *Bartholomew* (detail of Fig. 2.29) https://en.wikipedia.org/wiki/The_Last_Judgment_(Michelangelo)#/media/File:Michelangelo,_Giudizio_Universale_31.jpg; by Michelangelo—Web Gallery of Art: Image Info about artwork, Public Domain, https://commons.wikimedia.org/w/index.php?curid=11426515

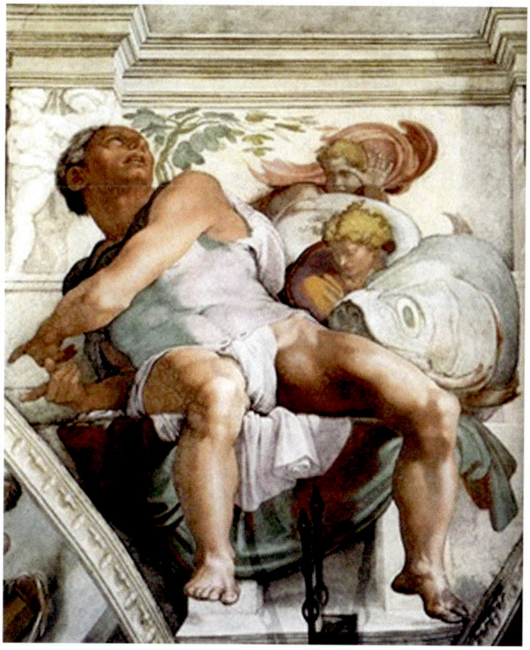

**Fig. 2.33** *Jonah* https://en.wikipedia.org/wiki/Prophet_Jonah_(Michelangelo)#/media/File:Prophet_Jonah.jpg; by Michelangelo–Michelangelo, Public Domain, https://commons.wikimedia.org/w/index.php?curid=957804

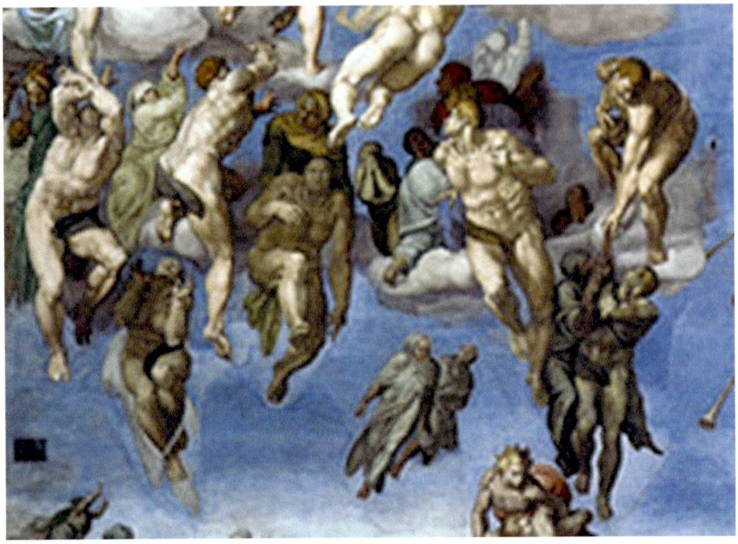

**Fig. 2.34** *Volterra's redactions* (detail of Fig. 2.29) https://en.wikipedia.org/wiki/The_Last_Judgment_(Michelangelo)#/media/File:Last_Judgement_(Michelangelo).jpg; by Michelangelo—see below. Public Domain, https://commons.wikimedia.org/w/index.php?curid=16143987. Cropped from original

In 1564 the Council of Trent decided that the nudity in the mural was inappropriate, and engaged the artist Daniele da Volterra (1509–1566) to paint over the genitalia. Volterra's crude handiwork is evidenced in several sections of the painting (Fig. 2.34). Michelangelo was so enraged that he broke into the Chapel and painted-in Baigio da Cesena, the Council Master, as Minos, the king of Hades, wearing goat's horns and being attacked rather harshly (though ironically) by a serpent (Fig. 2.35).[20]

[20] Further narrative and photos on this fascinating painting can be found at https://entertainment.howstuffworks.com/arts/artwork/michelangelos-last-judgment.htm#pt5; https://www.italian-renaissance-art.com/Last-Judgement.html; and http://www.artbible.info/art/large/54.html; http://www.thelastjudgement.org/Gallery.html.

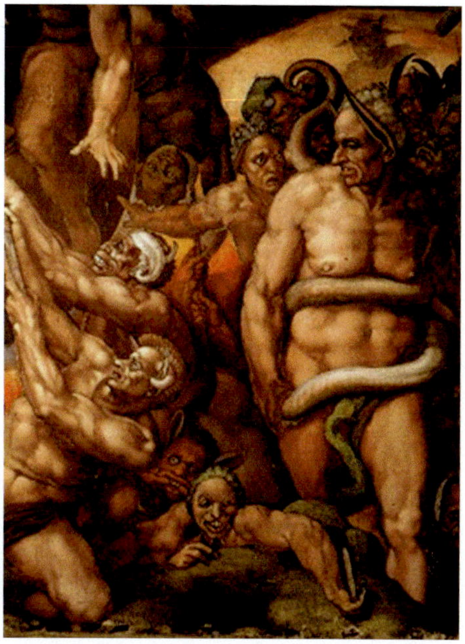

**Fig. 2.35** *Baigio da Cesena* (detail of Fig. 2.29) https://en.wikipedia.org/wiki/The_Last_Judgment_(Michelangelo)#/media/File:Michelangelo,_Giudizio_Universale_29.jpg; by Michelangelo—Web Gallery of Art: Image Info about artwork, Public Domain, https://commons.wikimedia.org/w/index.php?curid=11426510

Another interesting painting by Michelangelo is Leda and the Swan (Fig. 2.36), which rather graphically depicts the seduction (of Leda) by Zeus (taking the form of a swan). (The original work is lost.)

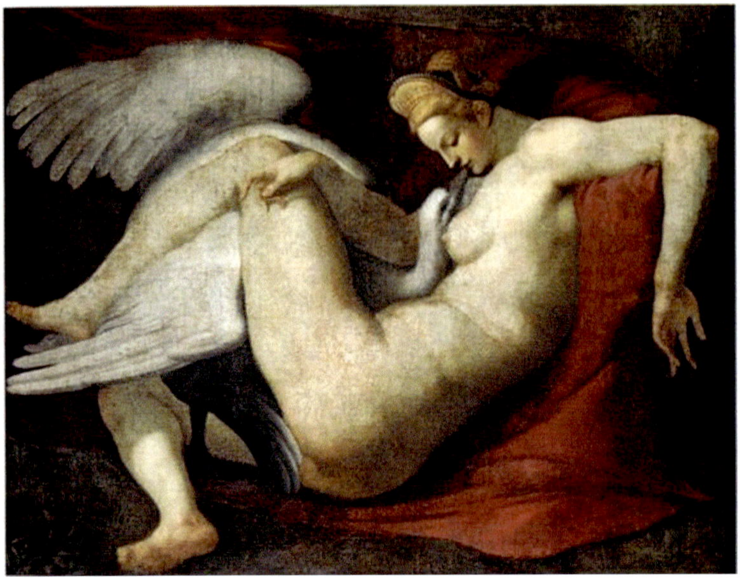

**Fig. 2.36** *Leda and the Swan* (**Michelangelo**) (copy by National Gallery, London)) by after Michelangelo—National Gallery, London—online catalogue. Public Domain, https://commons.wikimedia.org/w/index.php?curid=10107228. original 1530; copy, National Gallery, London

As a young engineer—just out of technical school, ignorant of art—I was fascinated by the demons depicted in The Last Judgement when I first saw it (Figs. 2.29 and 2.31). So this seems to be an appropriate moment for me to give Michelangelo a brief rest and indulge my inner child by sharing some intriguing examples of demonology in fine art.

## 2.6 Depictions of the Damned

Many artists, before and since Michelangelo, have exercised their imaginations in depicting Hades (professedly as a tool for encouraging conversion to Christianity). I have found these amusing, if a little unsettling—a puerile guilty pleasure—and I offer some samples without comment (Figs. 2.37, 2.38, 2.39, 2.40, 2.41, 2.42, 2.43, 2.44, 2.45, 2.46 and 2.47). Zoom in on the Bosch triptychs when you're feeling brave.

Figure 2.37 is a detail from the Last Judgement portal centering the west façade of Notre Dame cathedral in Paris. On the lintel you can see two demons trying to tip the scales as Archangel Michael weighs the merits of the newly risen dead; the fortunate souls are led to paradise on the left, while the damned are carted off to the tortures of hell on the right.

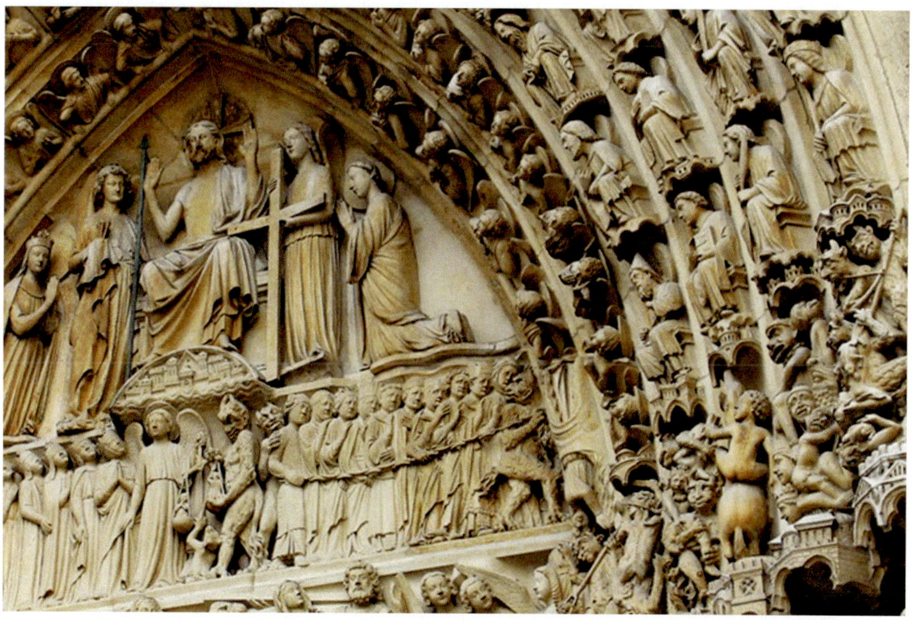

**Fig. 2.37** *Last Judgement* Photographed by Sarah Stierch from California—Notre Dame de Paris, CC BY 2.0, https://commons.wikimedia.org/w/index.php?curid=36730877. (Cropped from the original.) c. 1230; Notre Dame; Paris

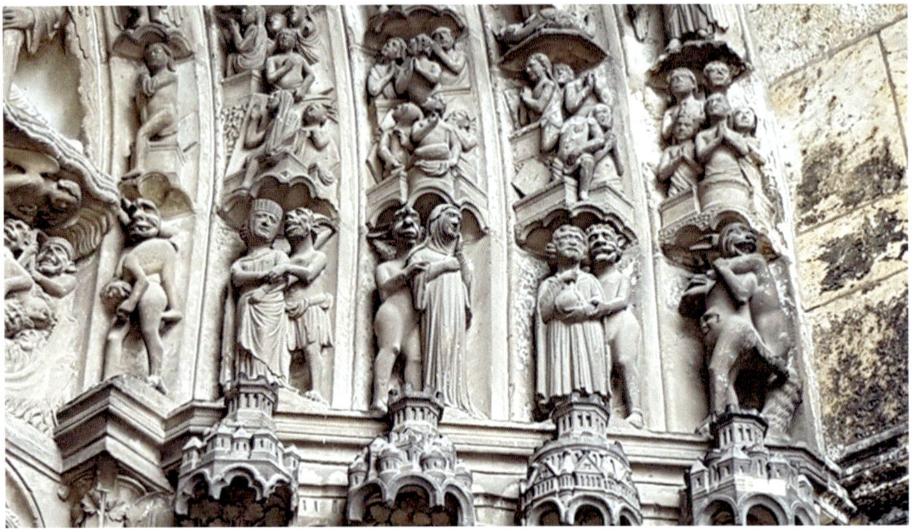

**Fig. 2.38** *Demons Tempting Christians* https://en.wikipedia.org/wiki/Chartres_Cathedral#/media/File:Chartres_-_south_portal_-_central_bay_-.jpg; by Seudo—own work, Public Domain, https://commons.wikimedia.org/w/index.php?curid=53163738. c. 1220; Chartres Cathedral, Chartres

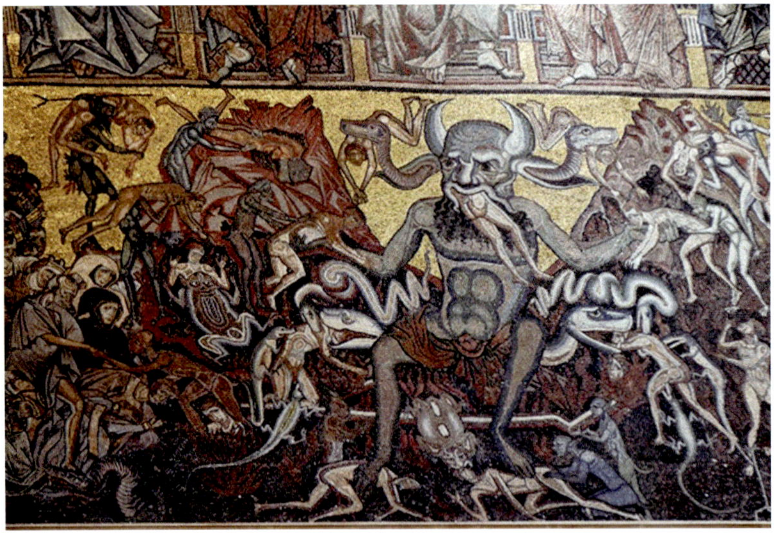

**Fig. 2.39** *Satan* (**Jacobus**) by © Marie-Lan Nguyen/Wikimedia Commons, CC BY 2.5, https://commons.wikimedia.org/w/index.php?curid=21571253. c. 1128; Baptistery of St. John; Florence

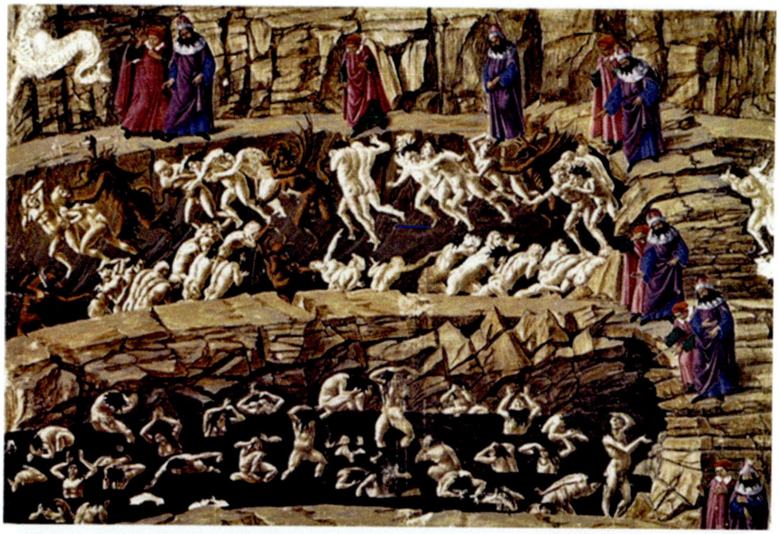

**Fig. 2.40** *8th circle of hell*, **Dante**'s *Divine comedy* (illustration by **Botticelli**) https://en.wikipedia.org/wiki/Divine_Comedy_Illustrated_by_Botticelli#/media/File:Botticelli_Inferno_XVIII.png; by Sandro Botticelli—jstor p. 182, Public Domain {{PD-US-expired-abroad}}, https://commons.wikimedia.org/w/index.php?curid=76621675

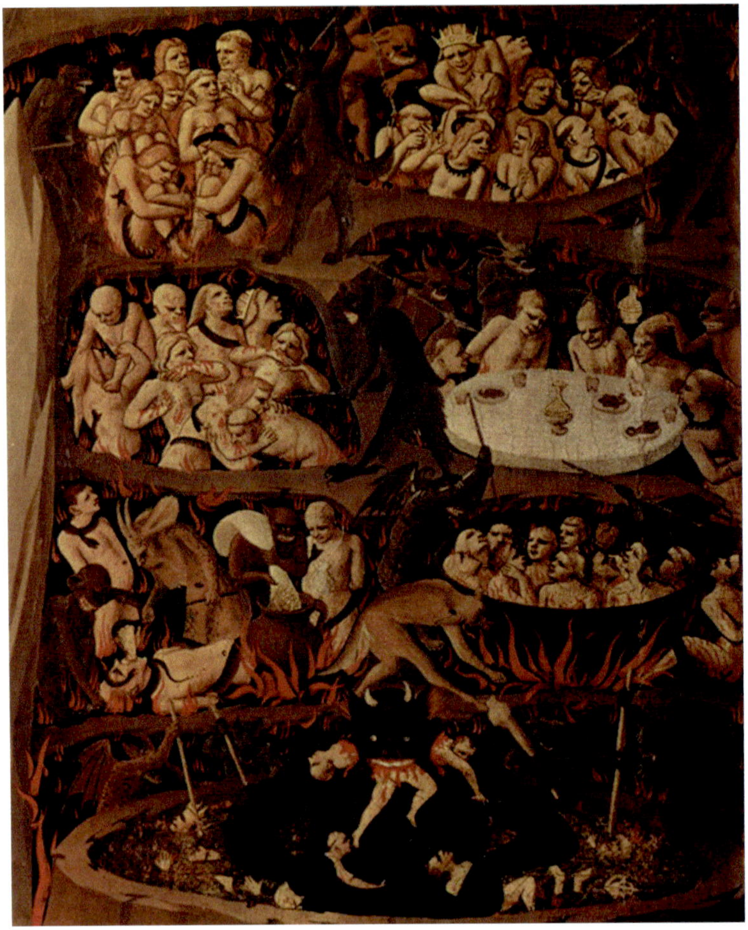

**Fig. 2.41** *Das Jungste Gericht* (Fra Angelico) The Yorck Project (2002) 10.000 Meisterwerke der Malerei (DVD-ROM), distributed by DIRECTMEDIA Publishing GmbH. ISBN: 3936122202. Public Domain {{PD-US}}, https://commons.wikimedia.org/w/index.php?curid=147478. 1435; Museum of San Marco, Florence

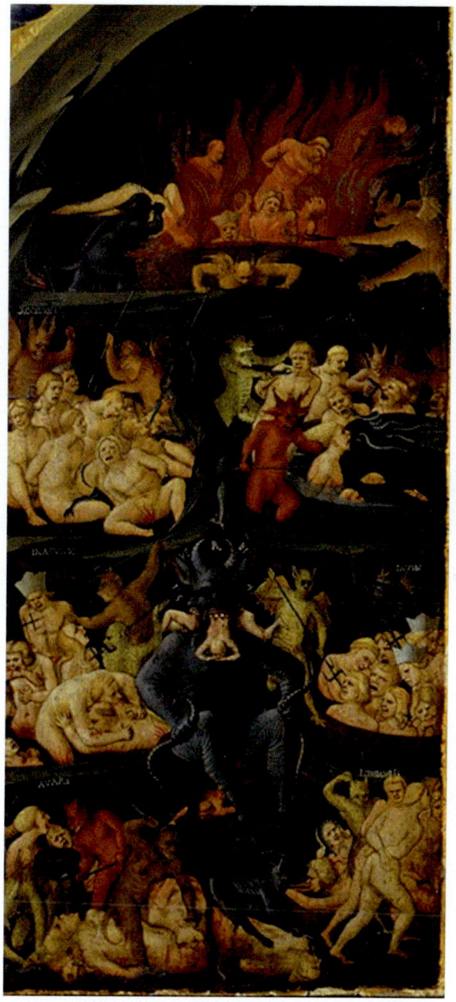

**Fig. 2.42** *Last judgement* (**Fra Angelica**) by Fra Angelico—EAEGi4U-_PXiIA—Google Arts and Culture, Public Domain {{PD-Pre 1978}}, https://commons.wikimedia.org/w/index.php?curid=13334067. (Cropped from original.) c. 1440; Gemäldegalerie, Berlin

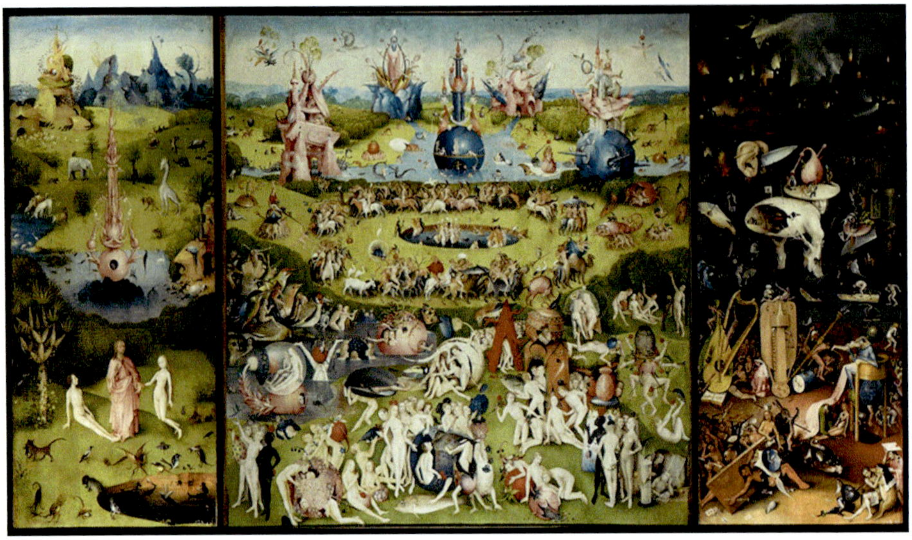

**Fig. 2.43** *Garden of Earthly delights* (**Hieronymous Bosch**) http://boschproject.org/dzi/00MCPVIS.dzi (downloaded with dezoomify-rs), Public Domain, https://commons.wikimedia.org/w/index.php?curid=90726084. c. 1505; Museo de Prado; Madrid

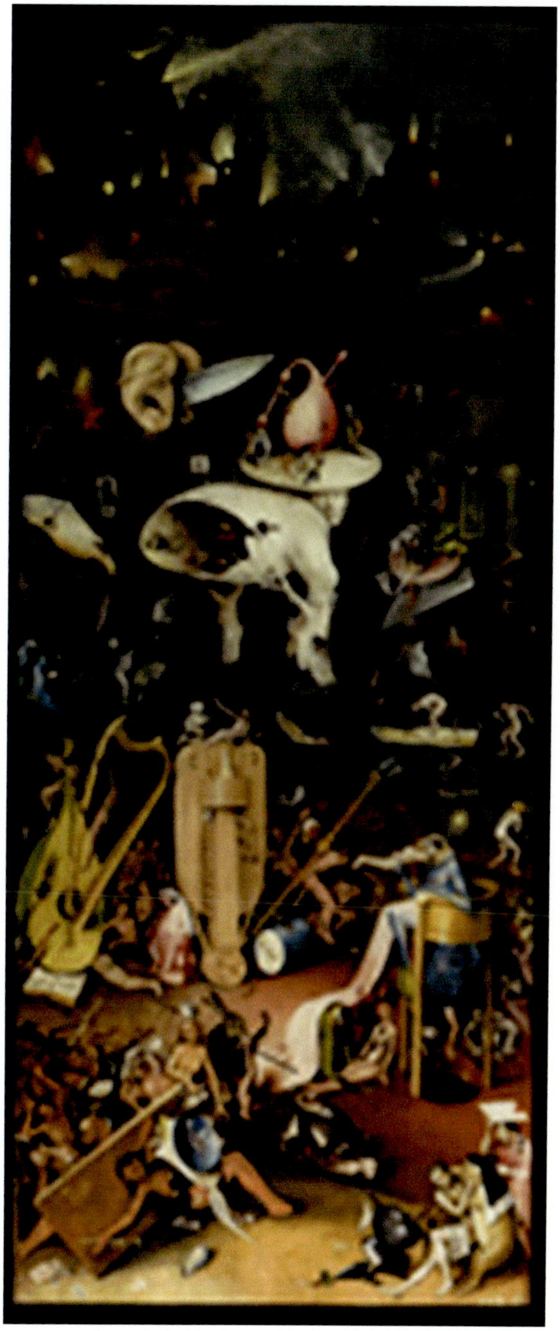

**Fig. 2.44** *Garden of Earthly delights* (detail; see Fig. 2.43) (cropped from original)

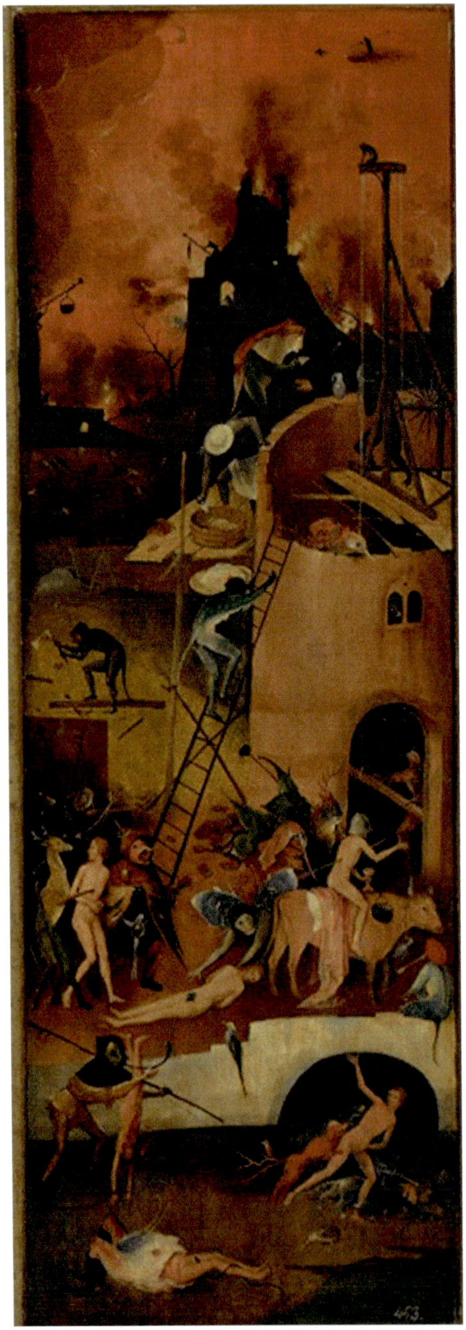

**Fig. 2.45** *Haywain* (**Bosch**) (detail) by Hieronymus Bosch or workshop—subscription art database, Public Domain {{PD-US}}, https://commons.wikimedia.org/w/index.php?curid=23521581. c. 1516; Museo del Prado, Madrid (cropped from original)

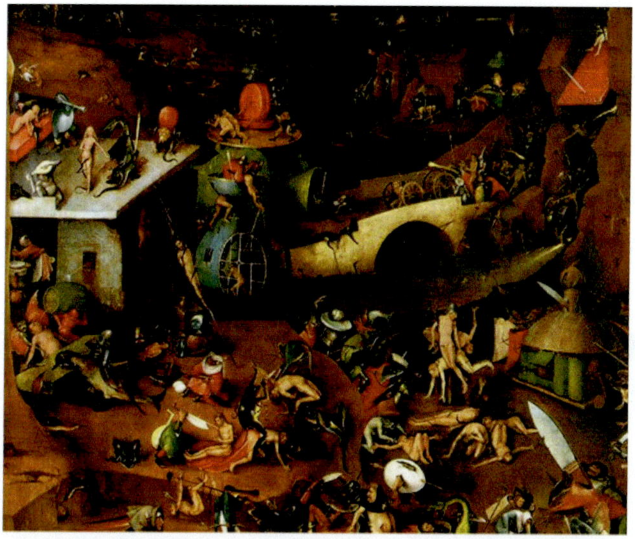

**Fig. 2.46** *Last judgement* (**Bosch**) (detail) by Hieronymus Bosch—www.statenvertaling.net: Home: Info, Public Domain, https://commons.wikimedia.org/w/index.php?curid=4593502. c. 1504; Academy of Fine Arts, Vienna (cropped from original)

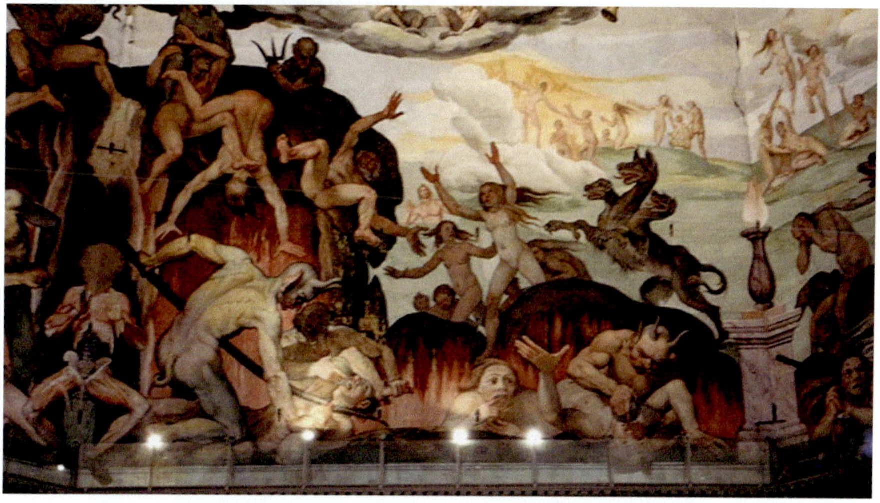

**Fig. 2.47** *Last judgement* (**Vasari** and **Zucarri**) (fresco detail) by Sailko—own work, CC BY 3.0, https://commons.wikimedia.org/w/index.php?curid=42491400. c. 1579; Cattedrale di Santa Maria del Fiore, Florence

## 2.7 Vasari

**Giorgio Vasari** (1511–1574) played an important role in art history. He is probably best known for authoring *Lives of the Most Excellent Painters, Sculptors, and Architects*,[21] the forerunner of art history books (Fig. 2.48). Its nearly 200 biographies contain, according to later historians, considerable inaccuracies and conjectures, but it is regarded as the most important work of Renaissance biography. No doubt it is the source of the story of Vasari's rescuing David's forearm.

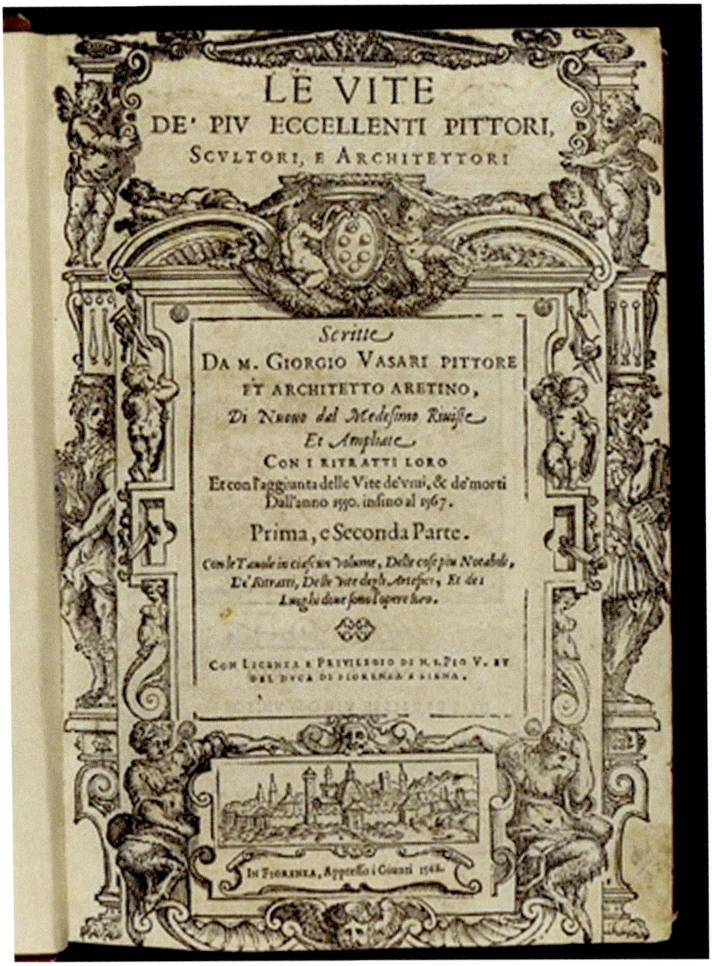


**Fig. 2.48** Cover of *The Lives* by **Giorgio Vasari**, Public Domain, https://commons.wikimedia.org/w/index.php?curid=3903220

[21] See https://www.google.com/search?client=firefox-b-1-d&q=Lives+of+the+Most+Excellent+Painters%2C+Sculptors%2C+and+Architects%2C+.

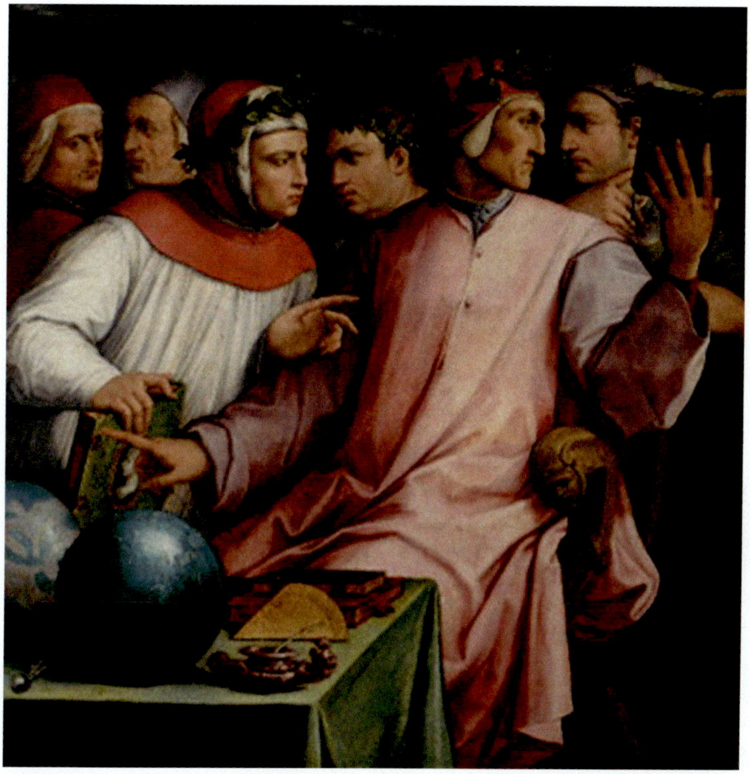

**Fig. 2.49** *Six Tuscan Poets* (**Giorgio Vasari**) Institute of Art Minneapolis Institute of Art—Google Arts and Culture, Public Domain, https://commons.wikimedia.org/w/index.php?curid=29675059. 1544; Minneapolis Institute of Art, Minnesota

Vasari studied painting with some of the masters in Florence, and was friends with (the older) Michaelangelo. His own paintings have, over the years, suffered in comparison with the masters. See if you don't agree with me that they lack a certain polish (Figs. 2.49 and 2.50).

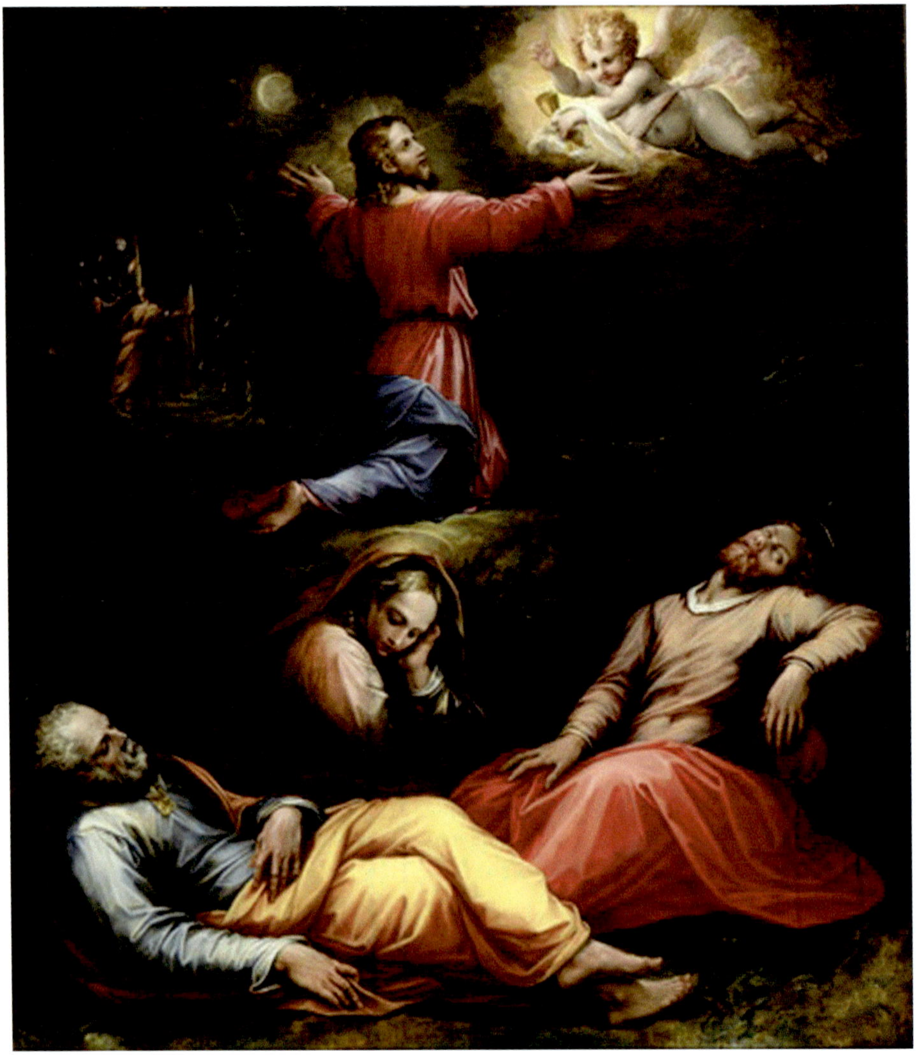

**Fig. 2.50** *Garden of Gethsemane* (**Vasari**) gAEsFmG2aloBJQ at Google Cultural Institute maximum zoom level, Public Domain, https://commons.wikimedia.org/w/index.php?curid=21897331. c. 1570; National Museum of Western Art, Tokyo

Indeed your author, a frequent visitor to Florence, has never seen a "coffee table book" devoted to Vasari's art.

It must be said that Vasari wasn't shy about depicting indelicate modes of punishment, on the ceilings and walls of Florence's buildings (Figs. 2.51 and 2.52).

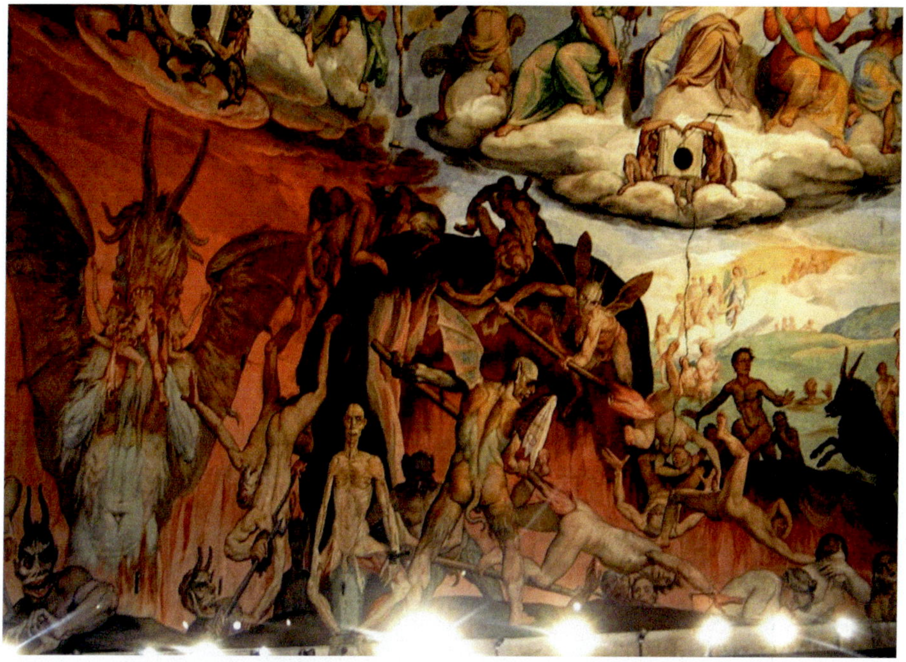

**Fig. 2.51** *Last judgement* (**Vasari** and **Zucarri**) (fresco detail) Photograph by Gareth Williams CC BY 2.0 Deed, https://www.flickr.com/photos/gareth1953/4060082530. c. 1579; Cattedrale di Santa Maria del Fiore, Florence

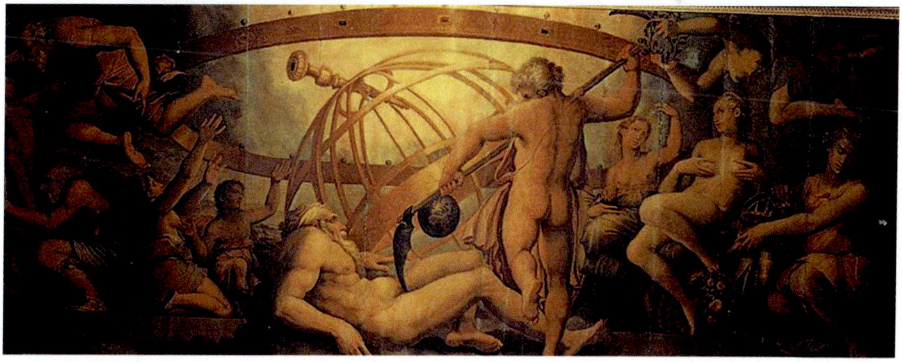

**Fig. 2.52** *Castration of Uranus* (**Vasari**, **Gheradi**)—http://www.uwm.edu/Course/mythology/0200/titans.htm, Public Domain {{PD-US}}, https://commons.wikimedia.org/w/index.php?curid=962858. c. sixteenth century; Palazzo Vecchio, Florence

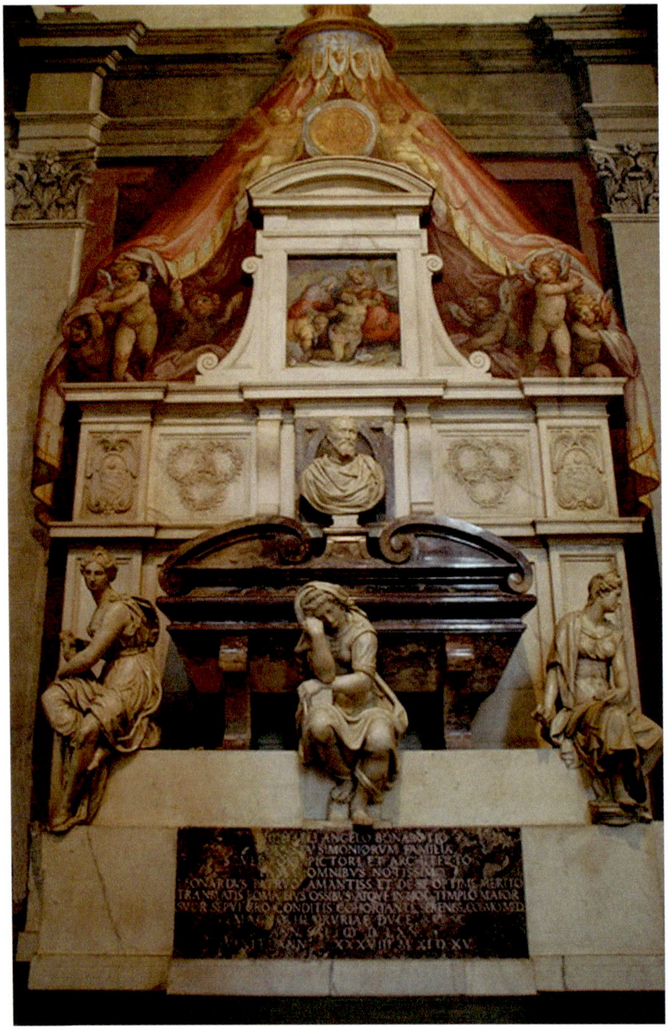

**Fig. 2.53** *Tomb of Michelangelo* (**Vasari**) Photo taken by Melissa Ranieri. Public Domain. https://en.wikipedia.org/w/index.php?curid=11654138. c. 1576; Santa Croce, Florence

However as an architect Vasari was a true Florentine hero. Tourists today still appreciate the Uffizi colonnade and loggia and his renovation of the church Santa Croce,[22] which houses his beautiful memorial for Michelangelo (as well as the memorials for Galileo, Dante, Macchiavelli, Marconi, and … Enrico Fermi!) (Fig. 2.53).

[22] https://en.wikipedia.org/wiki/vasari_corridor; https://en.wikipedia.org/wiki/Santa_Croce,_Florence.

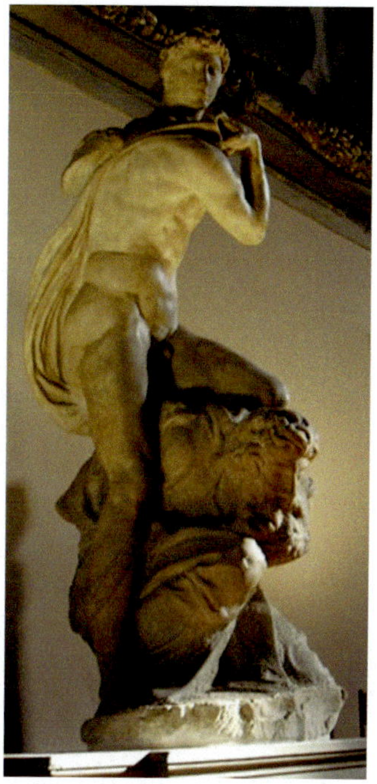

**Fig. 2.54** *Genius of Victory* (**Michelangelo**) by I. Sailko, CC BY 2.5, https://commons.wikimedia.org/w/index.php?curid=11399896. c. 1534; Salone dei Cinquecento, Palazzo Vecchio, Florence

In fact, Vasari's importance to the city of Florence has become even more evident recently. Inside Florence's Palazzo Vecchio there is a breathtaking assembly room known as the Salone dei Cinquecento[23] ("Hall of the Five Hundred"). It was featured in the 2001 movie *Hannibal*. The huge room is decorated with many murals—mostly by Vasari—and statues, including Michelangelo's *Genius of Victory* and Vincenzo di Rossi's *Hercules and Diomedes* (which failed to garner sanctioning by the World Wrestling Entertainment) (Figs. 2.54 and 2.55).

[23] https://en.wikipedia.org/wiki/palazzo_vecchio.

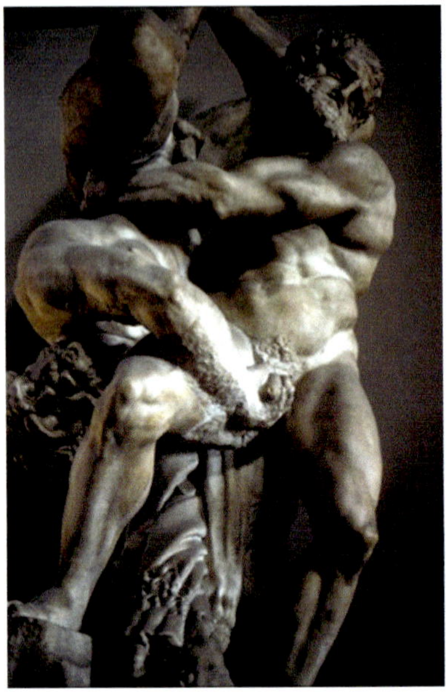

**Fig. 2.55** *Hercules and Diomedes* (**de Rossi**) Photo by Julian Stallabrass CC BY 2.0 DEED https://www.flickr.com/photos/slowkodachrome/. c. 1570; Salone dei Cinquecento, Palazzo Vecchio, Florence

But in 2012 the Salone featured in an intriguing news story.[24] It turns out that in 1504 Niccolò Machiavelli and Piero Soderini commissioned Leonardo da Vinci and the younger Michelangelo to paint murals on opposite walls of the Salone. Leonardo was supposed to depict the Battle of Anghiari and Michelangelo, the Battle of Cascina. There was definitely an element of competition; students of both artists stole and sabotaged each others' drafts. Michelangelo only completed a sketch ("cartoon") of his mural when the Pope recalled him to Rome to work on Julius's tomb.

It was believed that Leonardo actually painted his mural, but because he was experimenting with oils and wax the paint ran. His effort to salvage it by using candles to dry the paint faster only made it worse.

A few artists (notably **Peter Paul Reubens**) made copies of the unfinished works (Figs. 2.56 and 2.57), but around 1560 Grand Duke Cosimo I ordered Vasari to cover the walls and paint over them.[25]

[24] https://abcnews.go.com/International/leonardo-da-vincis-lost-painting-discovered-florence/story?id=15902573.

[25] https://en.wikipedia.org/wiki/palazzo_vecchio#salone_dei_cinquecento.

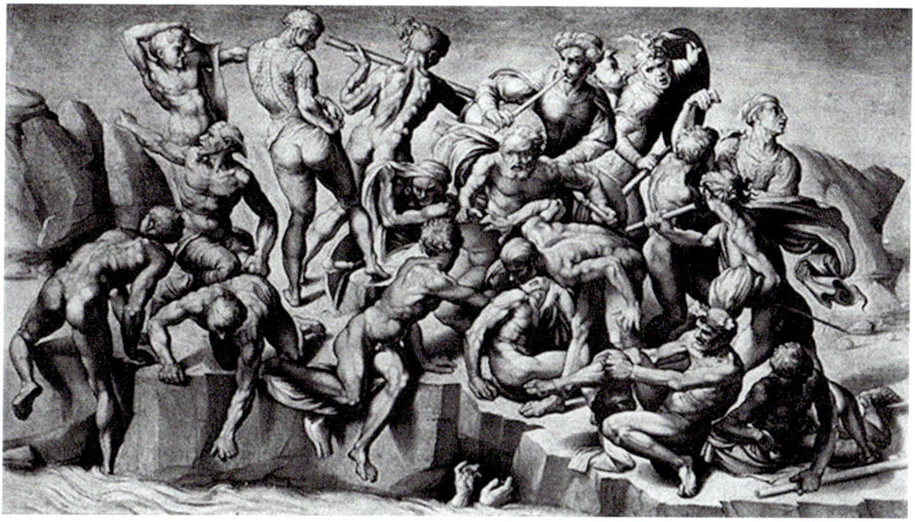

**Fig. 2.56** *The Battle of Cascina* (by **Bastiano da Sangallo** after **Michelangelo**)—http://it.wikipedia.org/wiki/File:Battagliadicascina.jpg, Public Domain, https://commons.wikimedia.org/w/index.php?curid=6760797. c. 1506; Holkham Hall, Holkam, Norfolk, England

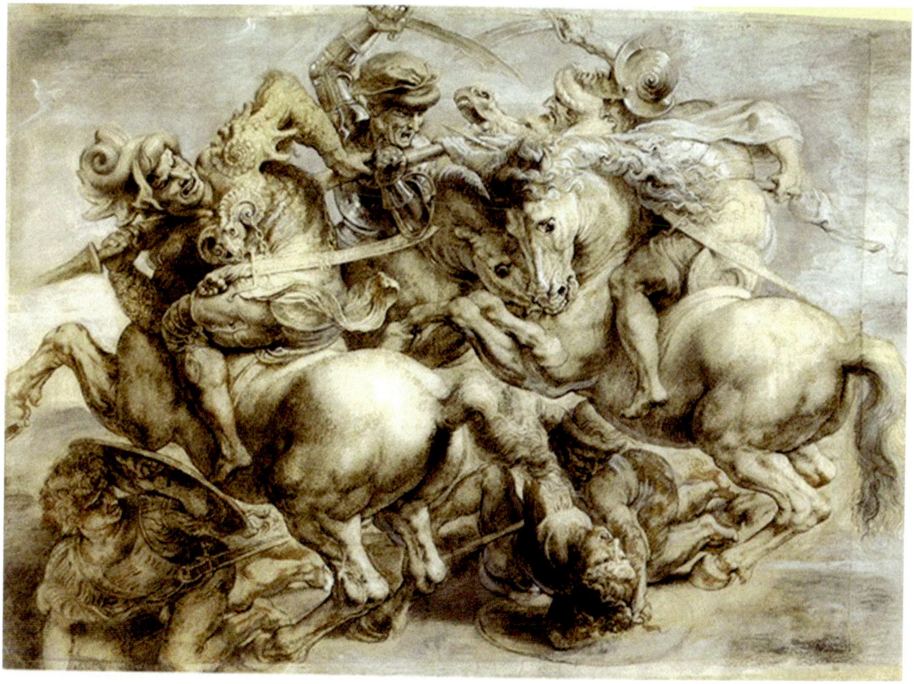

**Fig. 2.57** *The Battle of Anghiari* (by **Peter Paul Rubens** after **da Vinci)** Public Domain, https://commons.wikimedia.org/w/index.php?curid=59569. 1603; Louvre, Paris

Allusions to "the lost da Vinci" refer to the Battle of Anghiari. Various researchers, notably engineering Professor Maurice Seracini (who appeared in the movie *The da Vinci Code*) used existing technology to scan the walls of the Salone. He speculated that the Battle of Anghiari lies beneath Vasari's mural The Battle of Scannagallo.

**Exercise** Watch Seracini's fascinating 12-min talk[26] wherein he demonstrates some of the tools he used (laser scanning, thermography, radar, interferometry, ultrasound, fluorescence, X-rays, and even endoscopy—Seracini's undergraduate degree was in bioengineering) in his investigation.

Seracini concluded that our man Vasari did *not* overpaint the masterpiece; rather, he constructed a narrow brick wall in front of it, with a one-inch air gap, then put his own fresco on the brick wall—preserving the lost painting for posterity! Dr. Seracini drilled a few tiny holes through (restored sections of) Vasari's work and into the air gap, inserted flexible endoscope tubes, and ascertained that the adjacent wall does indeed contain oils and pigments commonly attributed exclusively to da Vinci. But "can one perforate a Vasari fresco in order to find a Leonardo painting?", as Stéphane Toussaint put it in his article "The Affair of the Battle of Anghiari" in *The Art Tribune*, June 13, 2012 (Paris: https://www.latribunedelart.com/). It's a sticky question,[27] and the Florentine museum officials suspended Seracini's research in 2012.[28]

An international conference[29] in 2016 reached the consensus that da Vinci never finished the painting, and the oils came from some other painting or the wall itself. As of 2020, Seracini advocated continuing the investigation.[30]

[26] https://www.ted.com/talks/maurizio_seracini_the_secret_lives_of_paintings; https://blog.ted.com/that-unicorn-is-really-a-lap-dog-the-secret-details-in-4-classic-paintings-revealed/.

[27] https://en.wikipedia.org/wiki/detachment_of_wall_paintings.

[28] https://www.huffpost.com/entry/da-vinci-battle-of-anghiari_n_1912719.

[29] R. Barsanti, G. Belli, E. Ferretti, C. Frosinini: *La Sala Grande di Palazzo Vecchio e la "Battaglia di Anghiari" di Leonardo da Vinci: dalla configurazione architettonica al all'apparato decorativo*. Biblioteca Leonardiana. Studi e Documenti, vol. 8. Olschki, Firenze (2019); A. Greenberger: Leonardo's 'Lost Masterpiece' can't be found because it was never painted, experts claim". https://www.artnews.com/art-news/news/leonardo-battle-of-anghiari-lost-masterpiece-art-history-1234573315/.

[30] *Leonardo's 'Lost Masterpiece'—Was It Ever Painted?* https://usaartnews.com/art-market/leonardo-s-lost-masterpiece-was-it-ever-painted; USA Art News (2020).

## 2.8 Michelangelo's Rivals

During the Renaissance, most of the world's best sculptors resided in Florence. Although photographs cannot do their works justice, here is a selection of my favorite pieces.

**Donatello** (Donato di Niccolò di Betto Bardi (1386–1466)) (Figs. 2.58 and 2.59).

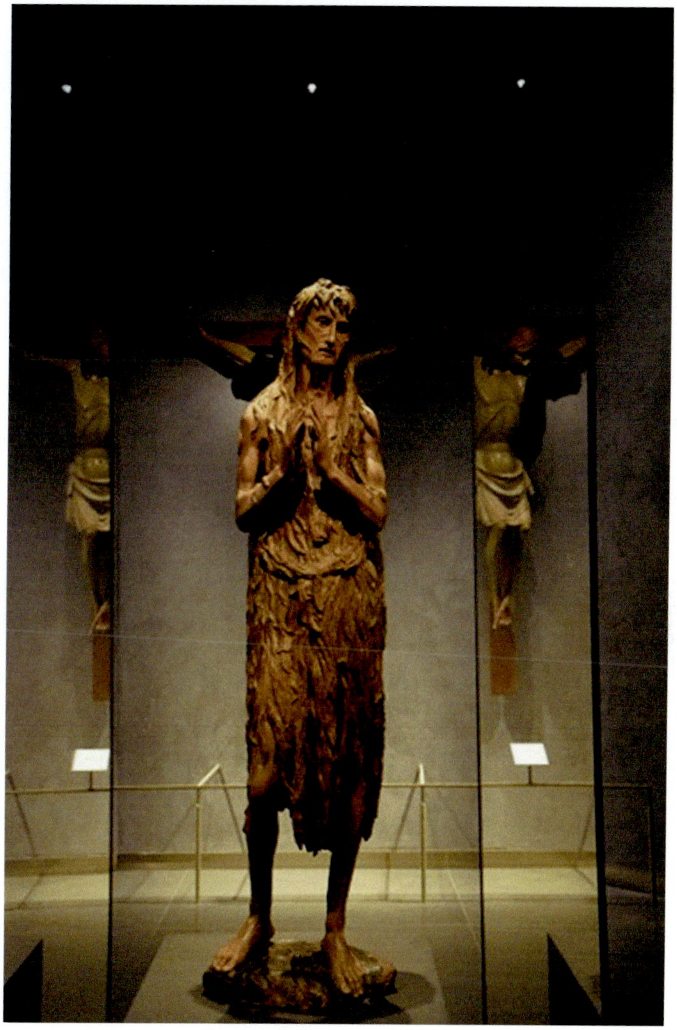

**Fig. 2.58** *Penitent Mary Maddalena* (**Donatello**) by George M. Groutas—https://www.flickr.com/photos/22083482@N03/48199247491/, CC BY 2.0, https://commons.wikimedia.org/w/index.php?curid=85727687. c. 1455; Museo dell'Opera del Duomo, Florence

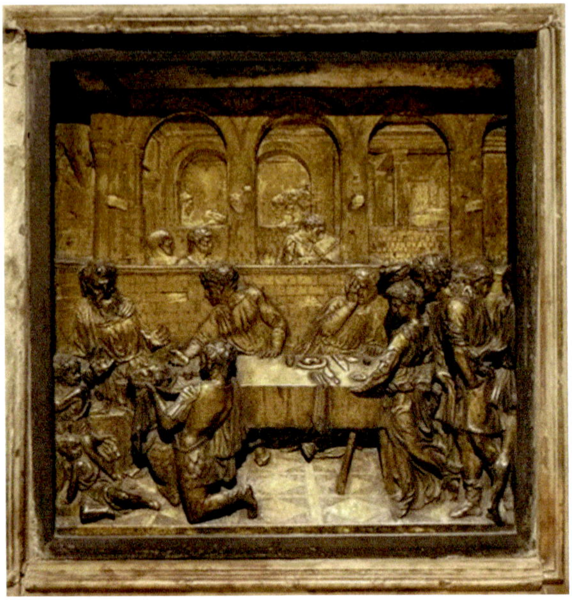

**Fig. 2.59** *Feast of Herod* (**Donatello**) by Donatello—own work, CC BY 3.0, https://commons.wikimedia.org/w/index.php?curid=44222907. 1420; Battistero di San Giovanni, Sienna

**Giambologna** (Jean Boulogne (1529–1608)) (Figs. 2.60, 2.61, 2.62 and 2.63).

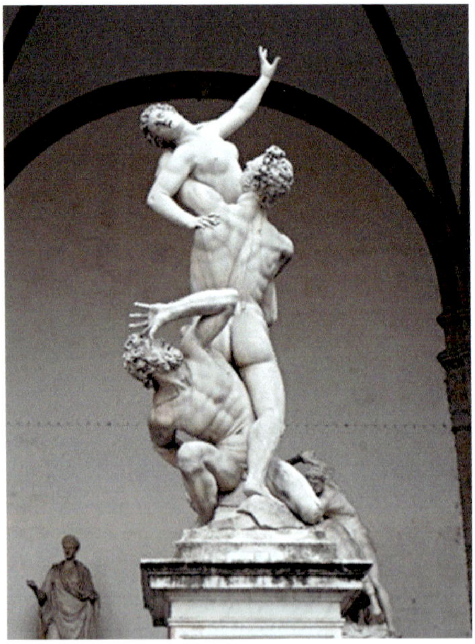

**Fig. 2.60** *Abduction of a Sabine Woman* (**Giambologna**) by Carlo Raso. Public Domain {{PDM 1.0}}. https://www.flickr.com/photos/70125105@N06/25064317965. 1583; Loggia della Signoria, Florence

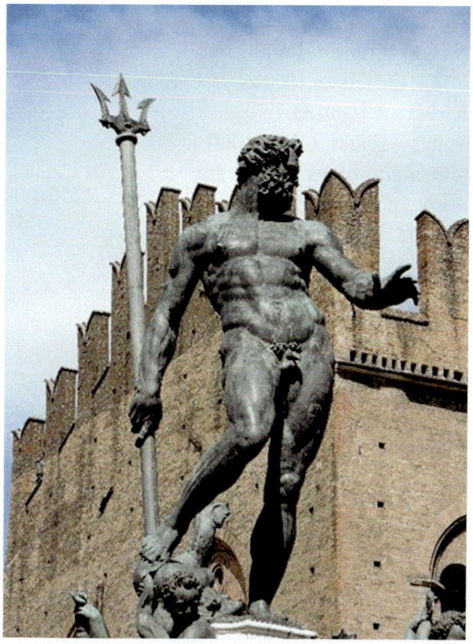

**Fig. 2.61** *Fountain of Neptune* (**Giambologna**) by Dimitris Kamaras from Athens, Greece—Fountain of Neptune (Fontana del Nettuno), Bologna, CC BY 2.0, https://commons.wikimedia.org/w/index.php?curid=118074729. c. 1567; Piazza del Nettuno, Bologna

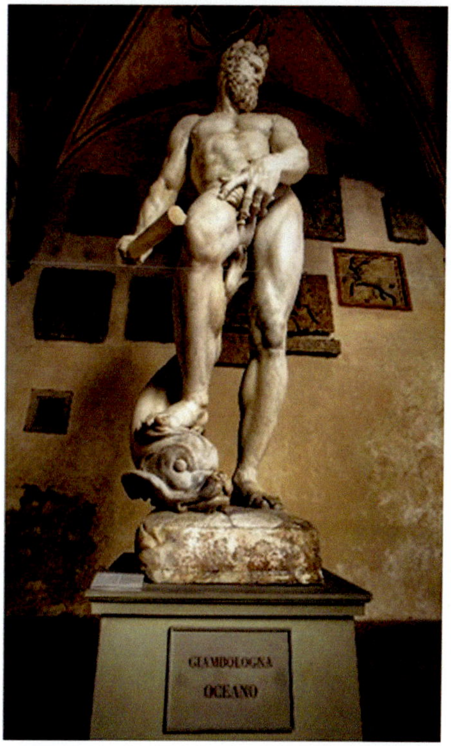


**Fig. 2.62** *Oceano* (**Giambologna**) Photo by aiva CC BY 2.0 DEED https://www.flickr.com/photos/aiva/. 1800; Museo Nazionale del Bargello, Florence

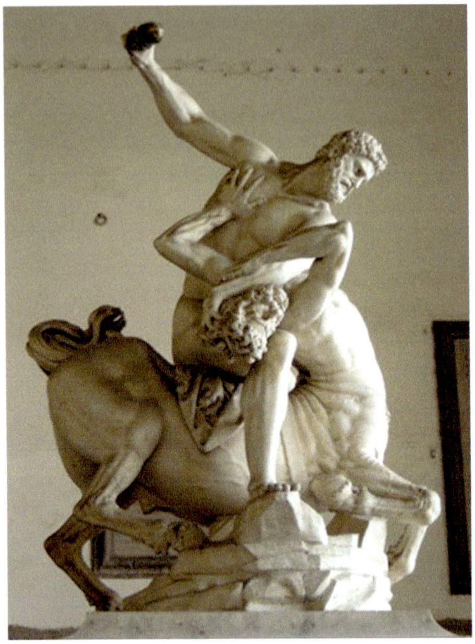

**Fig. 2.63** *Hercules and Nessus* (**Giambologna**) by Frank Fleschner (frankfl) from Kirksville, United States—Flickr, CC BY 2.0, https://commons.wikimedia.org/w/index.php?curid=275393. 1598; Piazza della Signoria, Florence

**Benvenuto Cellini** (1500–1571) (Fig. 2.64).

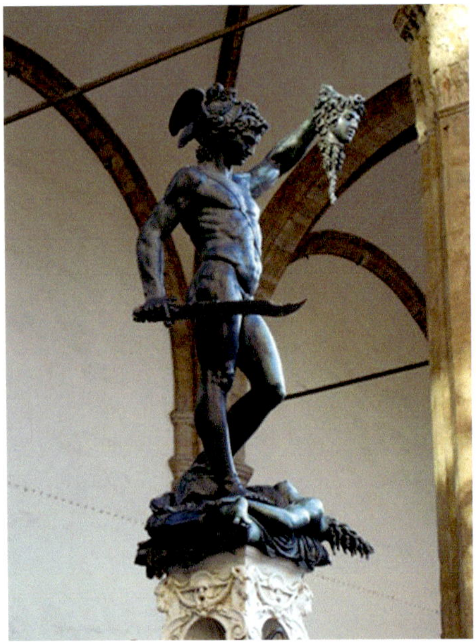

**Fig. 2.64** *Perseus with the Head of Medusa* (**Cellini**) by Jrousso at English Wikipedia—transferred from en.wikipedia to Commons. Public Domain, https://commons.wikimedia.org/w/index.php?curid=2374366. c. 1554; Piazza della Signoria, Florence

Most people recognize Michelangelo as the supreme sculptor. Indeed, the highest accolade another sculptor can achieve is to be compared to Michelangelo. This honor has been bequeathed to Rodin and Bernini, among others.

**Auguste Rodin** (1840–1917) (Figs. 2.65, 2.66, 2.67 and 2.68).

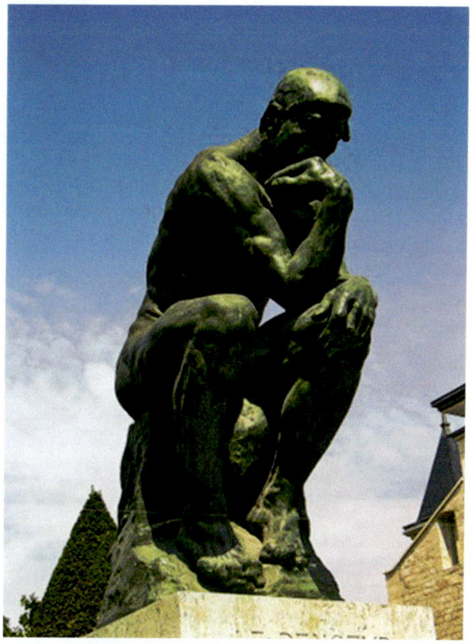

**Fig. 2.65** The Thinker (**Rodin**) by AndrewHorne (talk)—self-photographed (Original text: I (AndrewHorne (talk)) created this work entirely by myself.), Public Domain, https://commons.wikimedia.org/w/index.php?curid=15582363. 1904; Musee Rodin, Paris

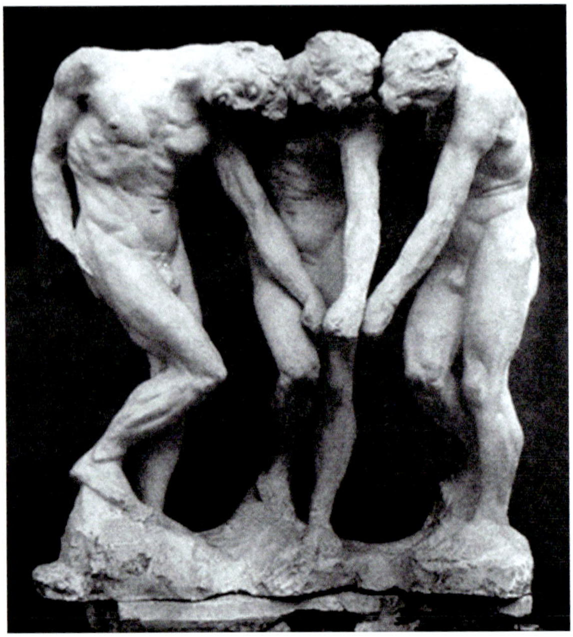

**Fig. 2.66** *The three shades* (**Rodin**) by unknown author—Burger, Fritz Cézanne und Hodler: Einführung in die Probleme der Malerei der Gegenwart, Munich: Delphin-verlag, Public Domain, https://commons.wikimedia.org/w/index.php?curid=26318764. c. 1886

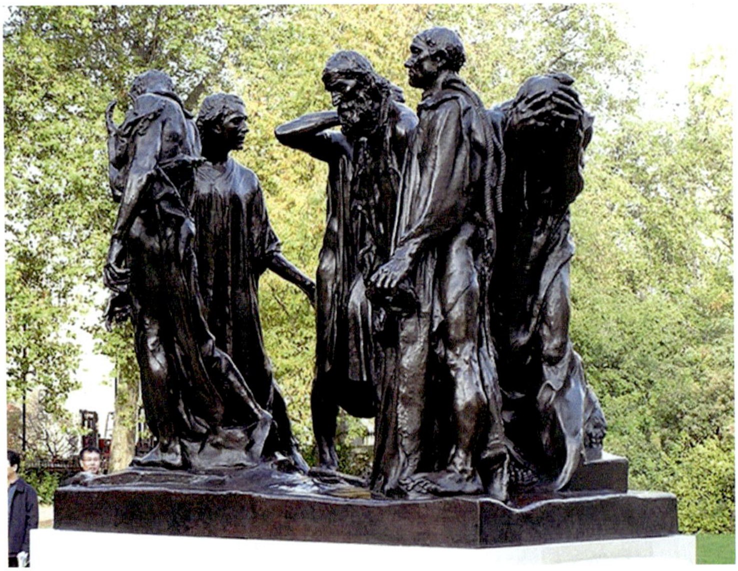

**Fig. 2.67** *The Burghers of Calais* (**Rodin**) Arpingstone at English Wikipedia, Public domain, via Wikimedia Commons https://commons.wikimedia.org/wiki/File:The.burg.of.calais.london.arp.750pix.jpg. c. 1889; Victoria Tower Gardens, London

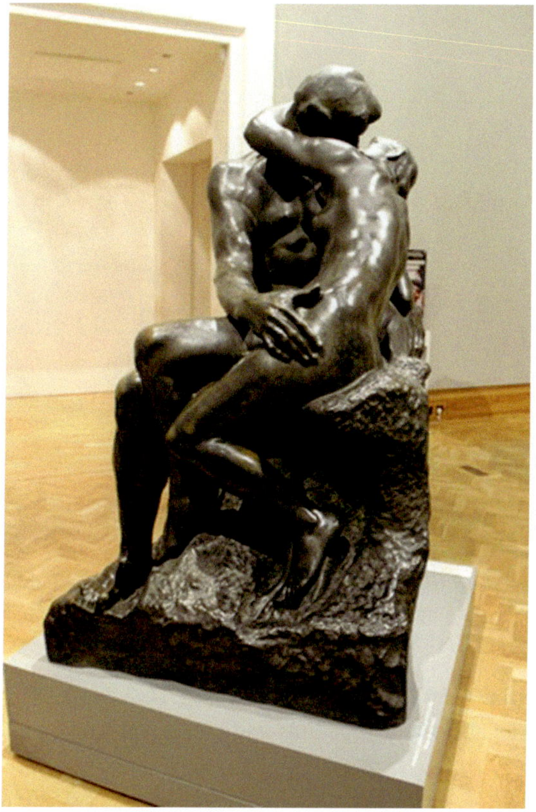

**Fig. 2.68** *The Kiss* (**Rodin**) 1887; by 14GTR—own work, CC0, https://commons.wikimedia.org/w/index.php?curid=141982275. c. 1887; National Museum, Cardiff

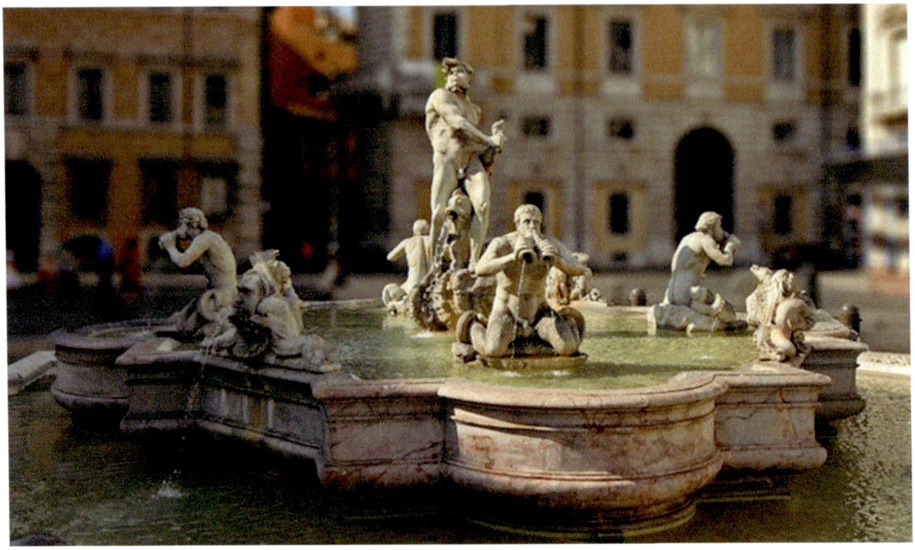

**Fig. 2.69** *Fontana del Moro* (**Bernini** et al.) by Jensens—own work, Public Domain, https://commons.wikimedia.org/w/index.php?curid=4584297. C. 1650; Piazza Navona, Rome

**Gian Lorenzo Bernini** (1598–1680) (Figs. 2.69, 2.70, 2.71, 2.72, 2.73, 2.74, 2.75 and 2.76). Bernini's works abound as fountains and piazzas in Rome and the Vatican:

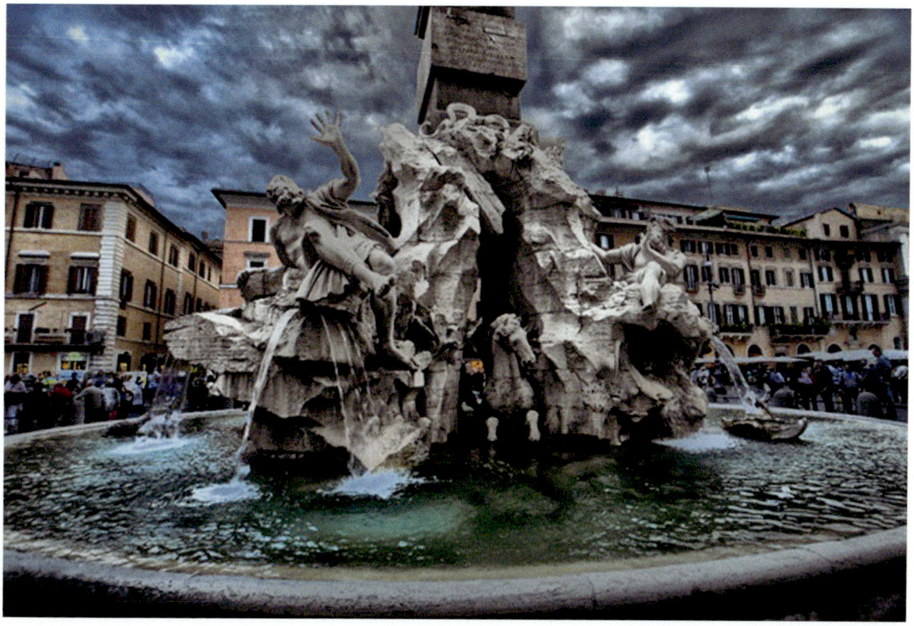

**Fig. 2.70** *Fontana dei Quattro Fiumi* (**Bernini**) Photo by Jose Luis Hidalgo R. CC BY 2.0 DEED https://www.flickr.com/photos/enfocalafoca/17328821974. 1651; Piazza Navona, Rome

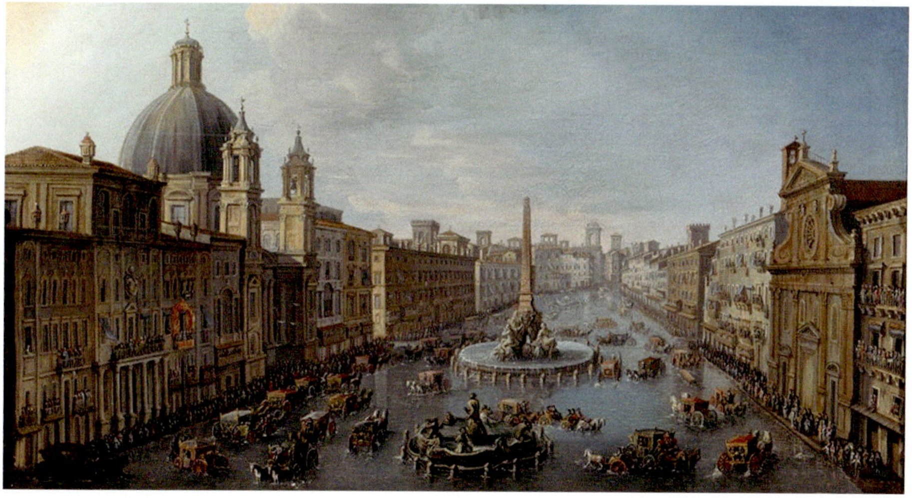

**Fig. 2.71** *View of the Flooded Piazza Navona* (**Antonio Joli**) Public Domain {{PD-US}}, https://commons.wikimedia.org/w/index.php?curid=101425023. Eighteenth century (foreground: *Fontana del Moro*; background: *Fontana dei Quattro Fiumi*)

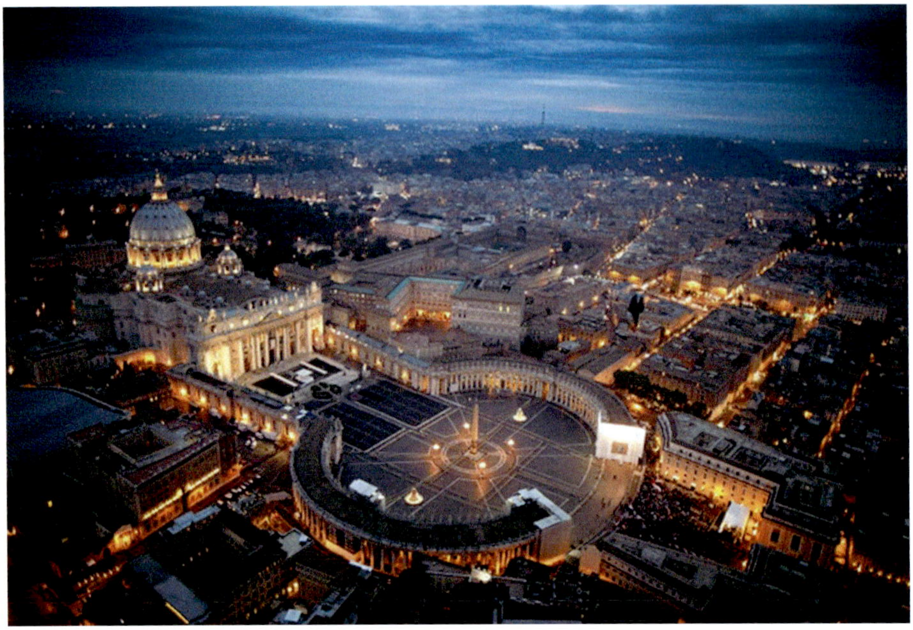

**Fig. 2.72** Saint Peter's Square (**Bernini**) by lafiguradelpadre Congreso—https://www.flickr.com/photos/lafiguradelpadre/8659426485, CC BY 2.0, https://commons.wikimedia.org/w/index.php?curid=77871816. c. 1667; Vatican City[31]

[31] A beautiful daytime photo can be found at https://commons.wikimedia.org/wiki/File:Collier%27s_1921_Vol_9_Frontispiece_--_Vatican.jpg.

But his sculptures are breathtaking. Look at *Apollo and Daphne* (Fig. 2.73).

Daphne treasures her virginity and asks the river god Ladon to transform her into a laurel tree lest Apollo violate her. The sculpture in Rome's Galleria Borghese hauntingly depicts Apollo's hand touching Daphne while parts of her body have already morphed into bark and leaves.

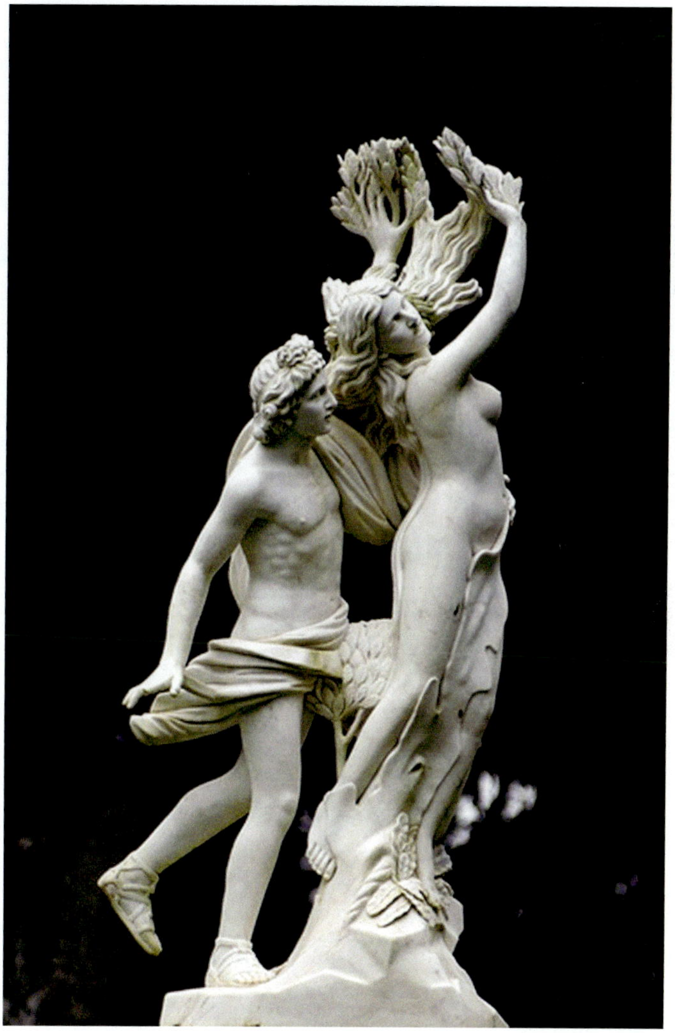

**Fig. 2.73** *Apollo and Daphne* (**Bernini**) by Guilherme Jofili After Gian Lorenzo Bernini, CC BY 2.0, https://commons.wikimedia.org/w/index.php?curid=16111540. c. 1625; Borghese, Rome (see a 3D model at https://sketchfab.com/3d-models/apollo-and-daphne-28f019be2e1143ec84af9ec7ca5be008)

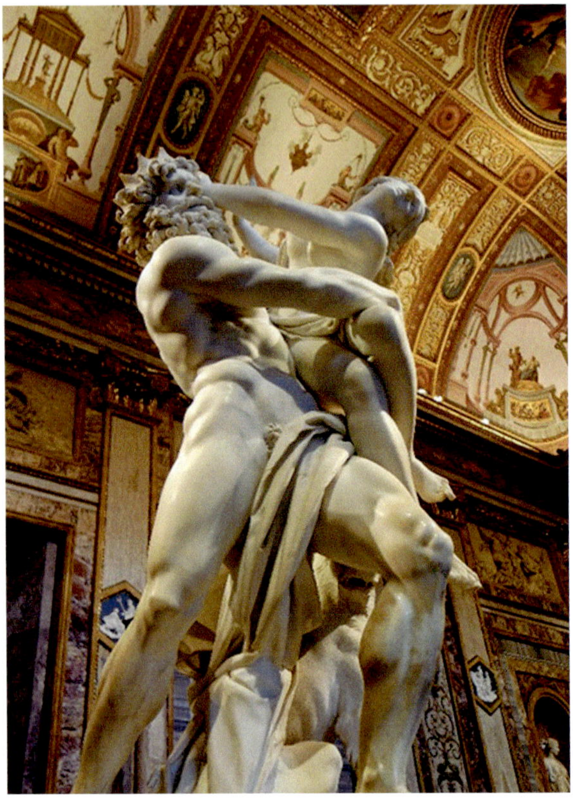

**Fig. 2.74** *Rape of Proserpina* (**Bernini**) by Sonse—The rape of Proserpina by Bernini, Galleria Borghese, CC BY 2.0, https://commons.wikimedia.org/w/index.php?curid=76176711. 1622; Borghese, Rome

In *Rape of Proserpina* notice how Bernini has skillfully carved the realistic indentations that Pluto's fingers make in Prosepina's flesh (Figs. 2.74, 2.75 and 2.76).

And when you have time take a brief look at some of my favorite American sculptures:

*Alice in Wonderland* (**Jose de Creeft** (1884–1982): https://en.wikipedia.org/wiki/File:Alice_in_Wonderland_sculpture_in_Central_Park.jpg).

*Abraham Lincoln Memorial* (**Daniel Chester French** (1850–1931): https://www.flickr.com/photos/publicresourceorg/493865318/in/photostream/).

*Detail* (**French**: https://www.flickr.com/photos/vintage_illustration/46252389091).

*Mount Rushmore* (**John Gutzon de la Mothe Borglum** (1867–1941): https://commons.wikimedia.org/w/index.php?curid=2098173).

*Seated Lincoln* (**Borglum**: https://commons.wikimedia.org/w/index.php?curid=10650867).

*Hans Christian Anderson* (**Georg John Lober** (1892–1961): https://commons.wikimedia.org/w/index.php?curid=64139668).

*George M. Cohan* (**Lober**: https://en.m.wikipedia.org/wiki/File:George_M._Cohan_by_Georg_Lober_-_DSC06436.JPG).

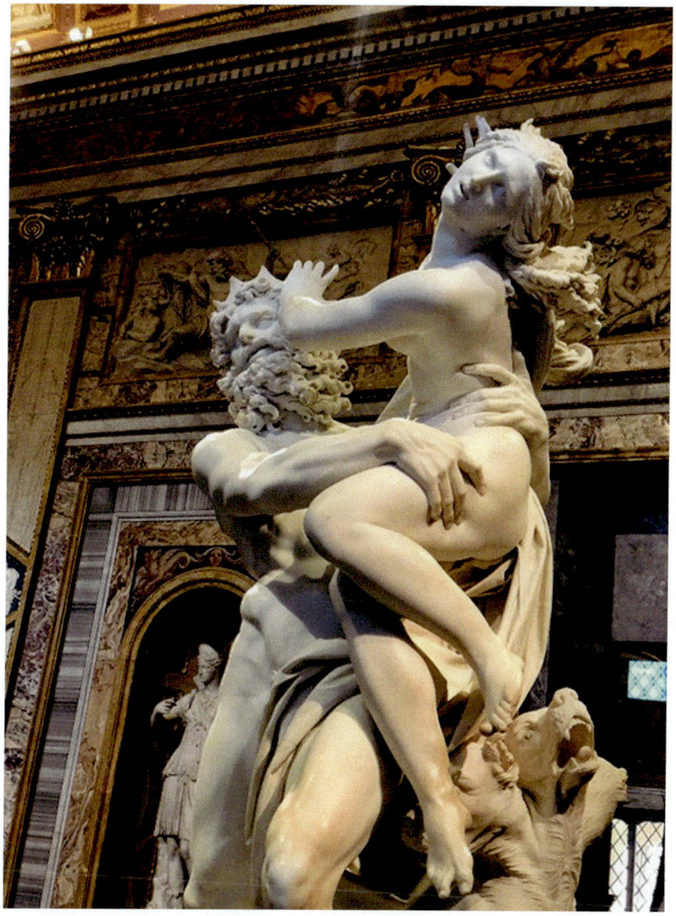

**Fig. 2.75** *Rape of Proserpina* (**Bernini**) by Sonse—The rape of Proserpina by Bernini, Galleria Borghese, CC BY 2.0, https://commons.wikimedia.org/w/index.php?curid=76175845. 1622; Borghese, Rome

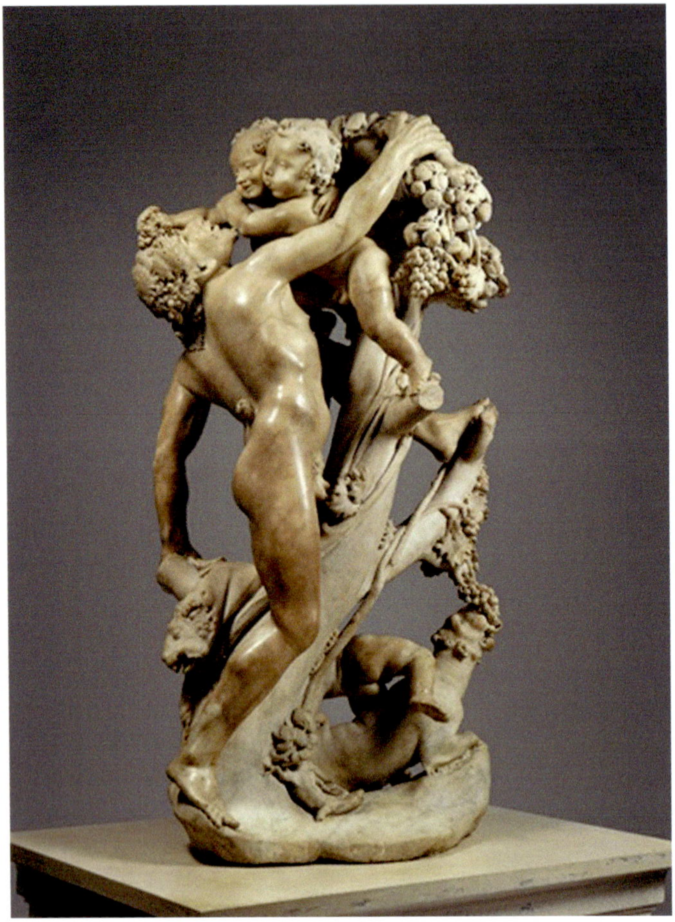

**Fig. 2.76** *Faun teased by children* (**Bernini**) https://www.metmuseum.org/art/collection/search/206399 CC0 1.0 https://en.m.wikipedia.org/wiki/File:Bacchanal-_A_Faun_Teased_by_Children_MET_DP248148.jpg 1616; Metropolitan Museum of Art, New York

As a painter, Michelangelo ranks among the three greats of renaissance art; in decreasing order of age: Leonardo, Michelangelo, and Rafael. Let's focus on Leonardo (Fig. 2.77).

**Leonardo da Vinci** (1452–1519)

Leonardo was the prototype for what we call the Renaissance Man. His own resume—immodest, as it should be—lists his skills as civil engineer, weapons designer, mechanical engineer, architect, sculptor, and painter. His sculpture has not survived. The most notorious work, a 23-ft bronze equestrian statue of the Duke of Milan, was never finished. Invading French soldiers used the clay model of the horse for target practice, and the bronze purchased for the casting had to be redirected for cannon construction. The *Sforza*

*Horse* in Milan (Fig. 2.78) was designed by **Nina Akamu** in 1999 from da Vinci's notes (https://en.wikipedia.org/wiki/Leonardo%27s_horse).

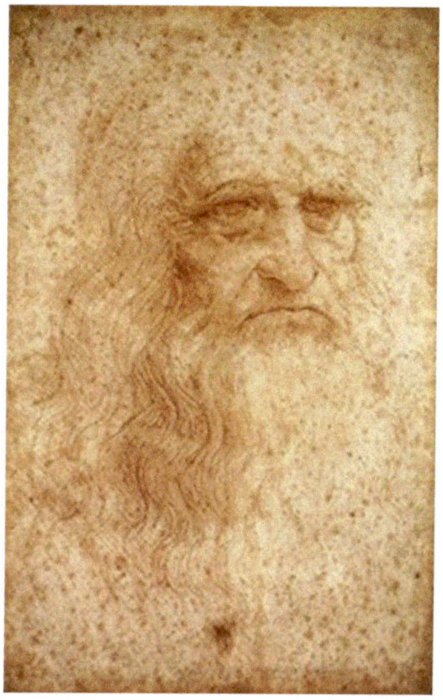

**Fig. 2.77** *Self portrait*(**da Vinci**) www.vivoscuola.it: Home; Picture, Public Domain, https://commons.wikimedia.org/w/index.php?curid=109273. c. 1518; Royal Library, Turin

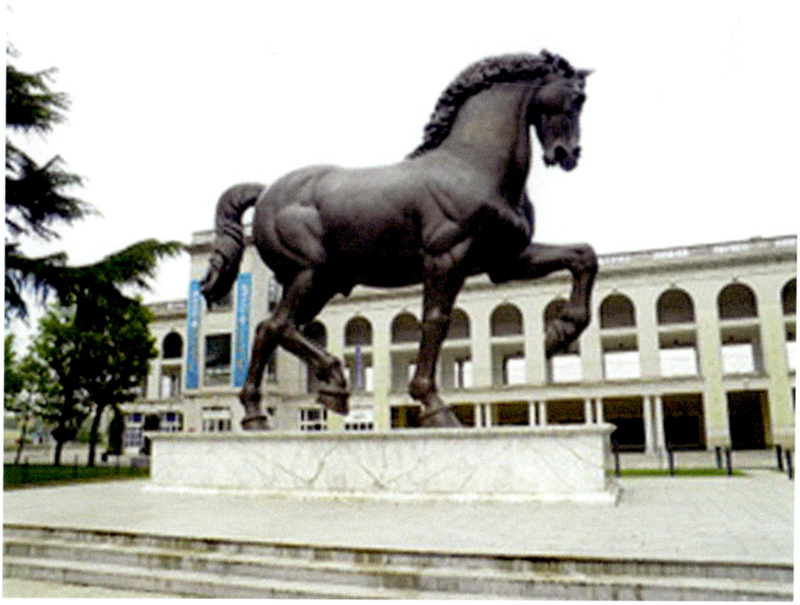

**Fig. 2.78** *Sforza Horse* (**Akamu**) by Mister No, CC BY 3.0, https://commons.wikimedia.org/w/index.php?curid=56308291. 1999; Hippodrome of San Siro, Milan

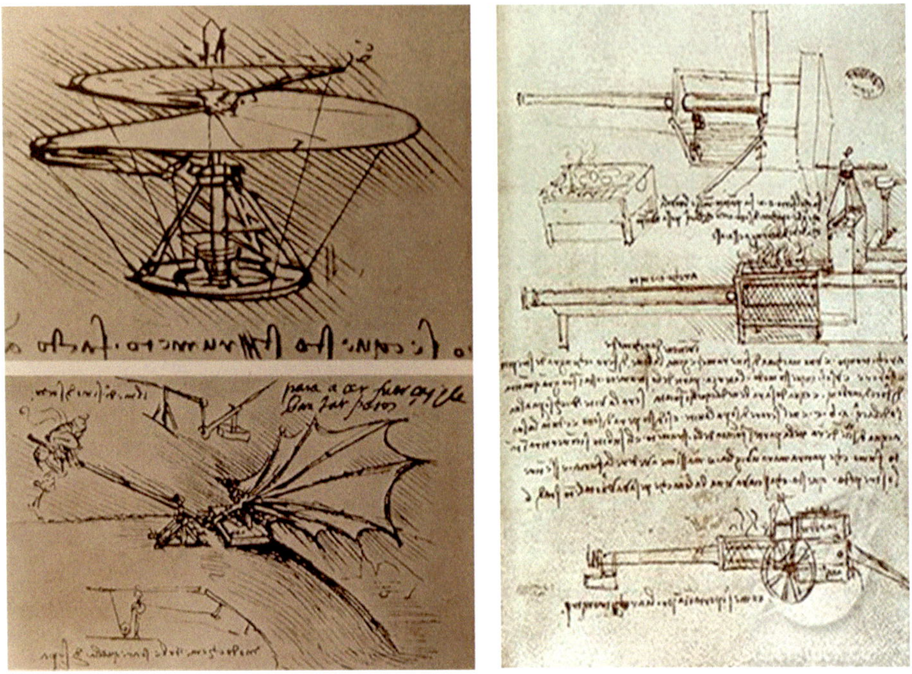

**Fig. 2.79** *Helicopter* (top left) *and experiment on lifting power of wing* (bottom left). (**da Vinci**) By Maksim. Public Domain, https://commons.wikimedia.org/w/index.php?curid=674315. c. 1480; British Museum, London. *Architonnerre* (right). (**da Vinci**)—http://www.superstock.com/stock-photos-images/463-4478, Public Domain {{PD-US}}, https://commons.wikimedia.org/w/index.php?curid=13299147. Fifteenth century

Most memorable among his invention/designs are his flying machines and cannons (Fig. 2.79).

His scientific observations were recorded in a mirror-reversed handwriting. To obtain a good knowledge of the human form, he dissected cadavers—often illegally (Fig. 2.80). (Michelangelo did the same.)

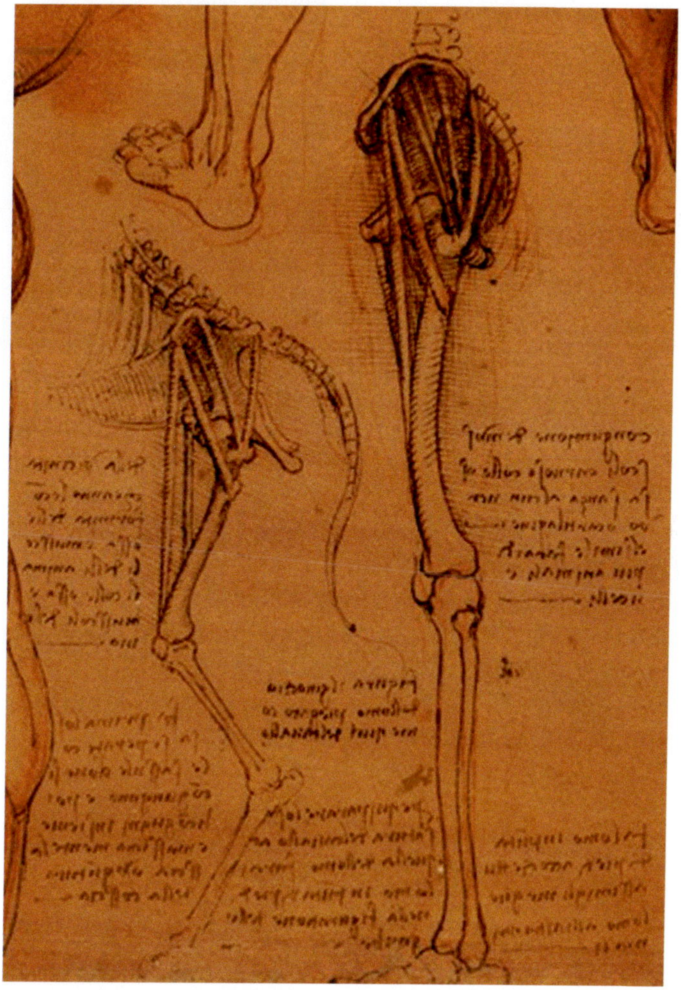

**Fig. 2.80** *Leg of dog and man* (**da Vinci**)—A.E.Popham "The Drawings of Leonardo", Public Domain {{PDArt}}, https://commons.wikimedia.org/w/index.php?curid=1647100. c. Sixteenth century

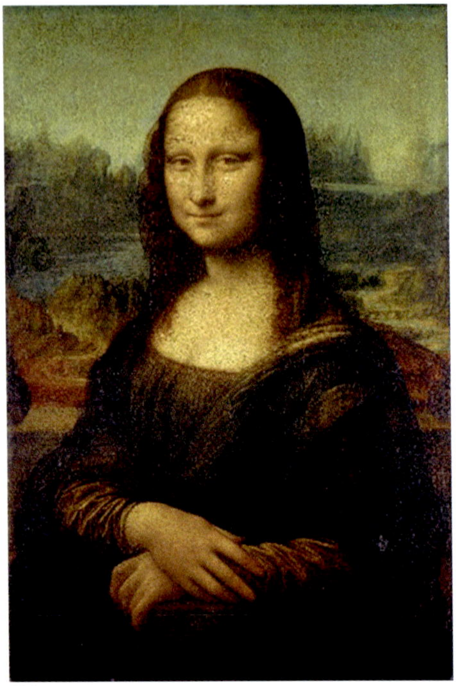

**Fig. 2.81** *Mona Lisa* (**da Vinci**) Cropped and relevelled from file: Mona Lisa, by Leonardo da Vinci, from C2RMF.jpg. Originally C2RMF: Galerie de tableaux en très haute définition: image page, Public Domain, https://commons.wikimedia.org/w/index.php?curid=15442524. c. 1506; Louvre, Paris

But we know da Vinci best by his paintings—most significantly, the best-recognized painting in the world, *Mona Lisa* (Fig. 2.81). She has resided at the Louvre museum in Paris since 1797, except for two years (1911–1913) when she was stolen. She has suffered acid burns and a thrown rock (1956, 2009) and being pied (2022), but bullet-proof glass has shielded her from vandals since 1957.

Other beloved works by Leonardo include *The Last Supper*, *Adoration of the Magi*, and *Vitruvian Man* (Figs. 2.82, 2.83 and 2.84). We'll have more to say about some of these in Chapter Three, when we discuss perspective.

You will enjoy reading *Leonardo and the Last Supper,* Ross King, Bloomsbury USA (2013), ISBN-10: 1620403080, ISBN-13: 978-1620403082.

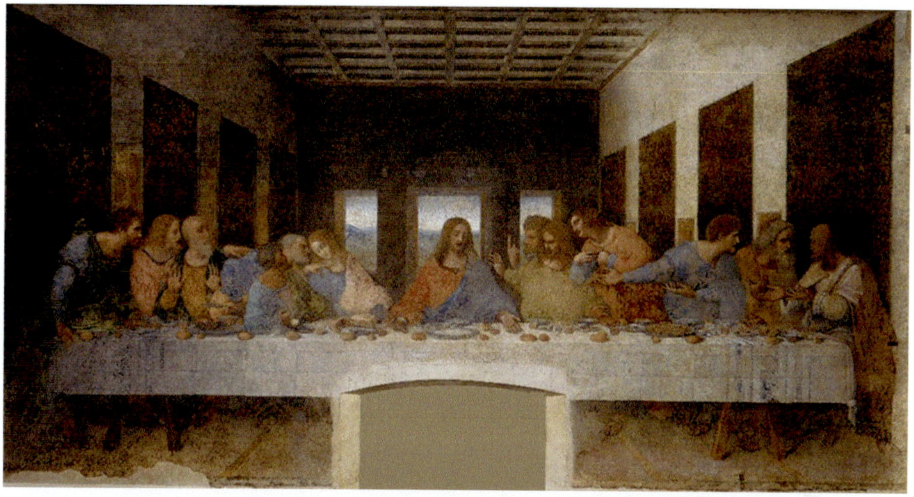

**Fig. 2.82** *The Last Supper* (**da Vinci**)—unknown source, Public Domain, https://commons.wikimedia.org/w/index.php?curid=24759. c. 1498; Dining Hall Wall in Santa Maria delle Grazie Convent, Milan

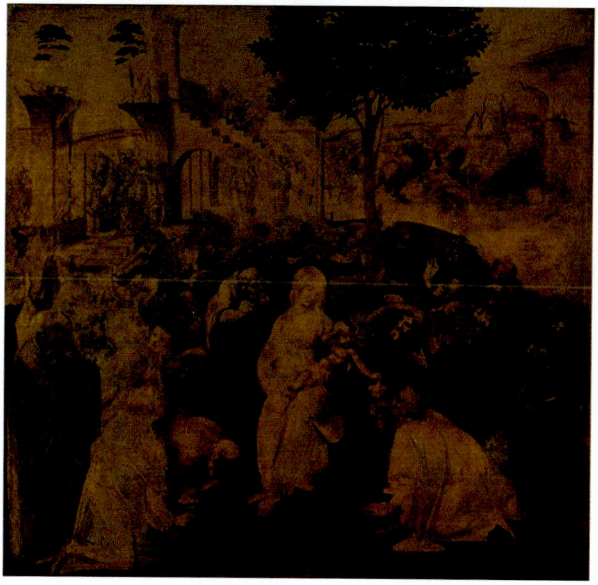

**Fig. 2.83** *Adoration of the Magi* (**Leonardo da Vinci**)—RQFL5tibYCPGOg—Google Arts and Culture, Public Domain, https://commons.wikimedia.org/w/index.php?curid=13514343. (unfinished) c. 1482; Uffizi Gallery, Florence

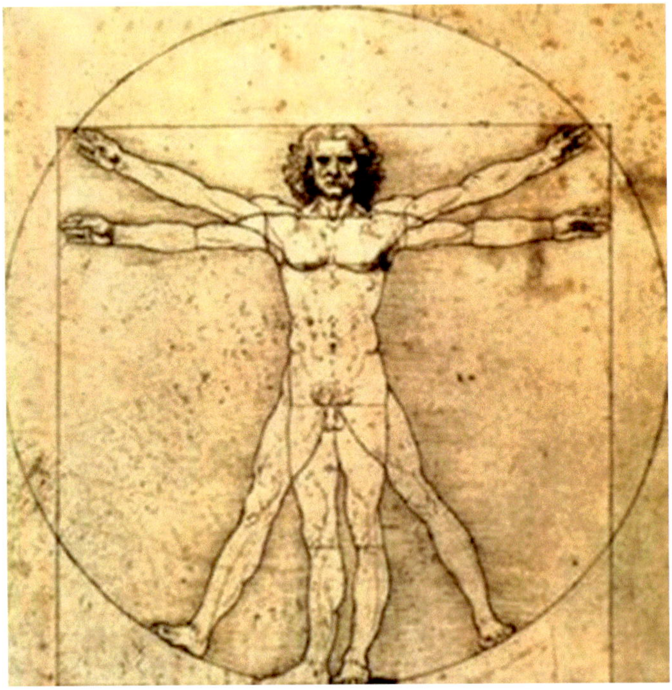

**Fig. 2.84** *Vitruvian Man* (**da Vinci**) Photo by Ortuns Public Domain https://en.m.wikipedia.org/wiki/File:Homme_de_vitruve.jpg. c. 1492; Gallerie dell'Accademie, Florence

**Raphael** (Raffaello Sanzio da Urbino, 1483–1520)

The novelist Dan Brown (*The da Vinci Code, Angels and Demons*) considers Raphael as the most skilled painter of the Great Three. His best-known works adorn the walls of the Raphael Rooms in the Vatican (Figs. 2.85, 2.86, 2.87 and 2.88). The School of Athens is breathtakingly beautiful and geometrically precise; it contains images of Leonardo, Michelangelo, and Rafael. (Enlarge the image and you'll be able to find them. Consult https://en.wikipedia.org/wiki/The_School_of_Athens for guidance.)

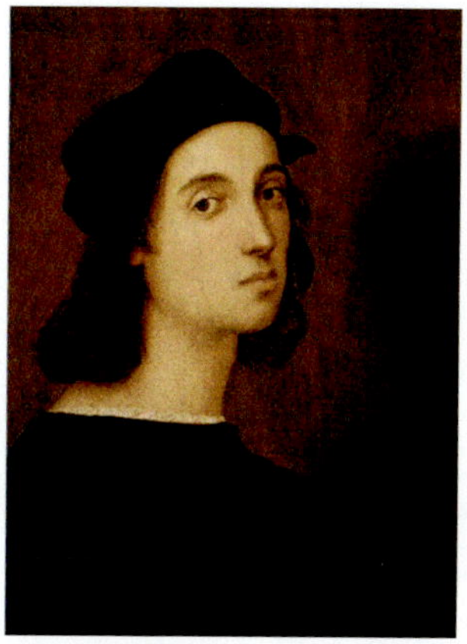

**Fig. 2.85** *Self portrait* (**Raphael**)—http://nevsepic.com.ua/art-i-risovanaya-grafika/2409-raffaello-sanzio-rafael-santi-37-rabot.html image, Public Domain, https://commons.wikimedia.org/w/index.php?curid=35440864. c. 1506; Pallazo Pitti, Florence

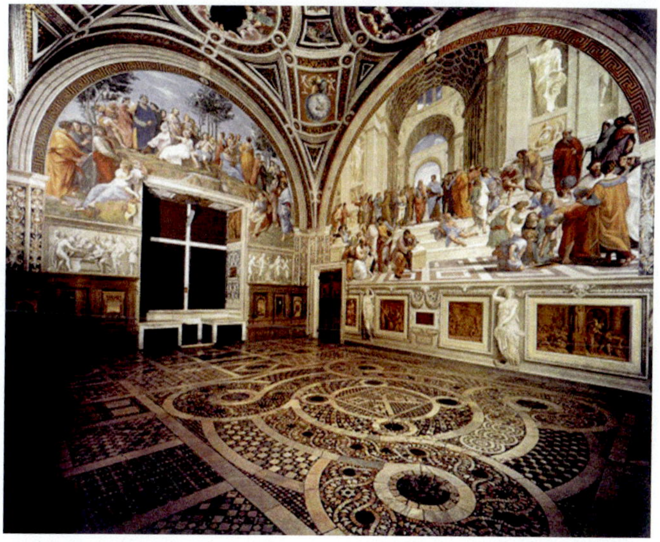

**Fig. 2.86** *Stanza della Segnatura* (**Raphael**) by Lure—own work, Public Domain, https://commons.wikimedia.org/w/index.php?curid=8534727. c. 1511; Raphael Rooms, Apostolic Palace, Vatican City

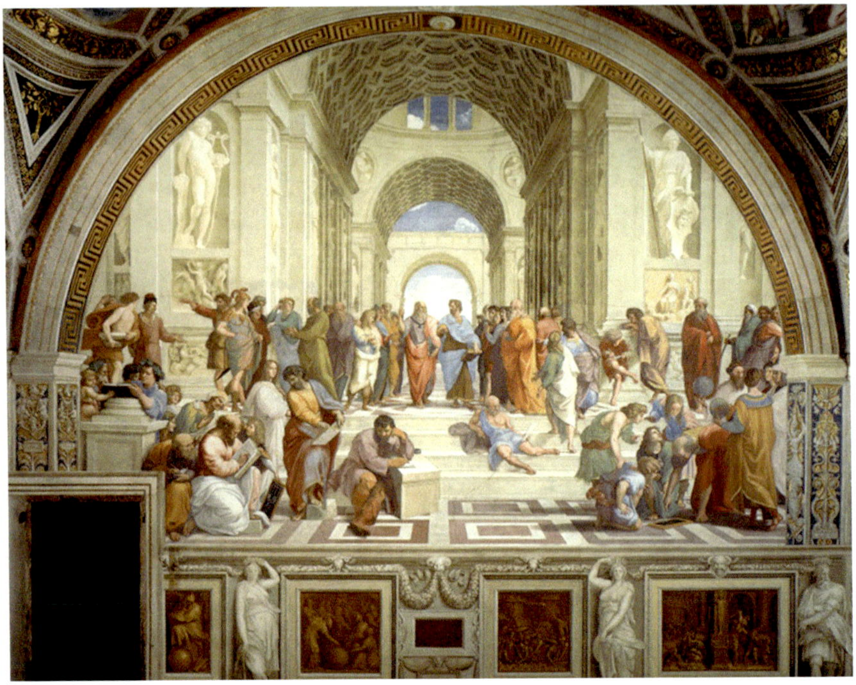

**Fig. 2.87** *School of Athens* (**Raphael**)—stitched together from vatican.va, Public Domain, https://commons.wikimedia.org/w/index.php?curid=4406048. c. 1511; Raphael Rooms, Apostolic Palace, Vatican City

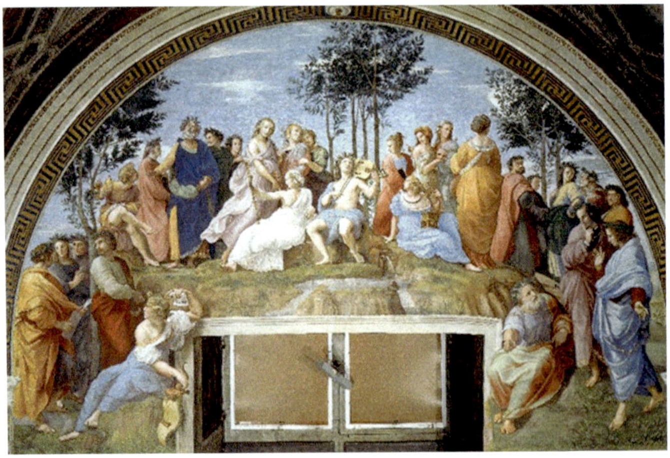

**Fig. 2.88** *Parnassus* (**Raphael**)—extracted from Parnaso (Raffaello), Ennio Dante e Omero.jpg, Public Domain {{US-PD}}, https://commons.wikimedia.org/w/index.php?curid=15709581. 1511; Vatican Museums, Vatican

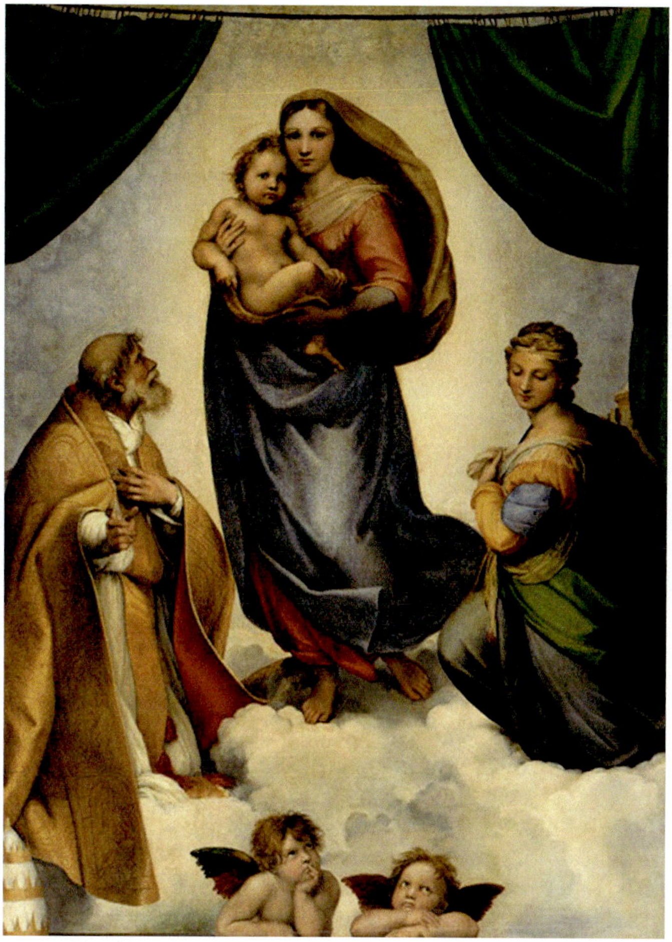

**Fig. 2.89** *Sistine Madonna* (**Raphael**)—Google Art Project: Home—pic Maximum resolution. Public Domain {{PD-US}}, https://commons.wikimedia.org/w/index.php?curid=19991622. c. 1513; Staatliche Kunstsammlungen Dresden, Dresden

And who doesn't recognize Raphael's cherubs from every card shop in the mall (Figs. 2.89 and 2.90)?[32]

[32] (The adorable little *bastardi* were reputedly the children of his model.)

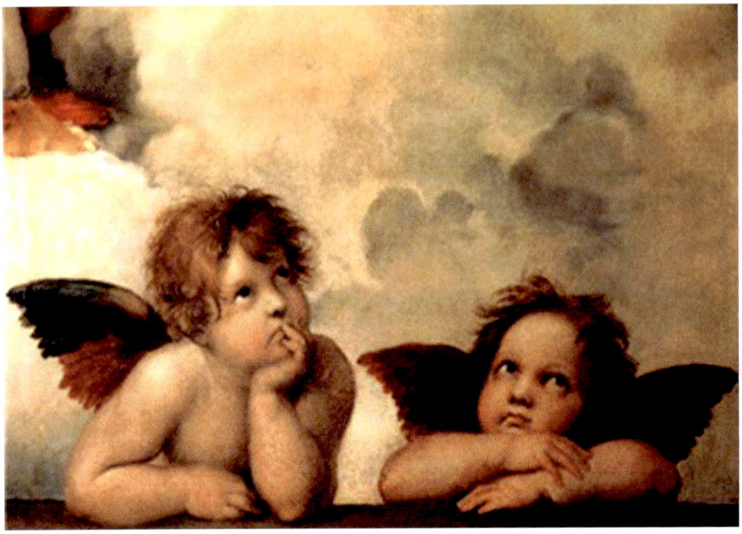

**Fig. 2.90** (*Detail of Fig. 2.89*) (**Raphael**)—[1], Public Domain {{PD-US}}, https://commons.wikimedia.org/w/index.php?curid=10634971

# 3 Perspective

## 3.1 Examples of Perspective

Take a look at the medieval paintings in Figs. 3.1 and 3.2.[1]

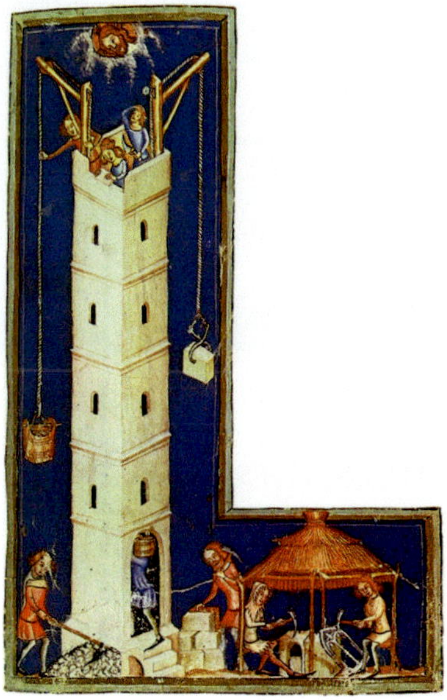

**Fig. 3.1** *Tower of Babel* (**Meister der Weltenchronik**) The Yorck Project (2002) *10.000 Meisterwerke der Malerei* (DVD-ROM), distributed by DIRECTMEDIA Publishing GmbH. ISBN: 3936122202. Public Domain 1370; Munich

[1] Creative Commons License Links for all images are listed in the Preface.



A. D. Snider, *From STEM to STEAM*, Synthesis Lectures on Engineering, Science, and Technology, https://doi.org/10.1007/978-3-031-57316-3_3

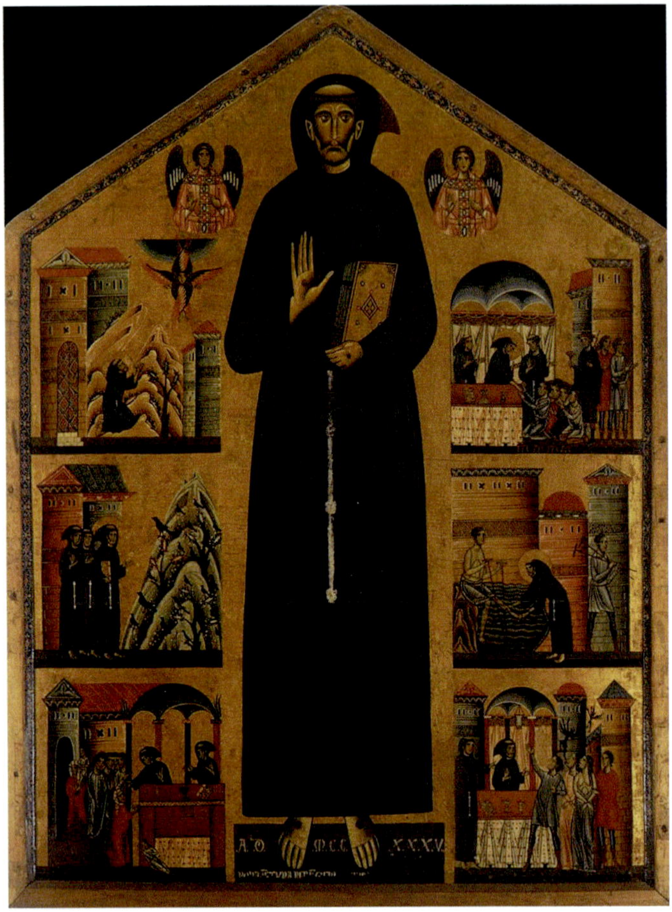

**Fig. 3.2** *Saint Francis of Assisi* (**Berlinghieri**)—Web Gallery of Art: Image Info about artwork, Public Domain{{PD-US}}, https://commons.wikimedia.org/w/index.php?curid=158831781235. 1235; Church of St. Francis

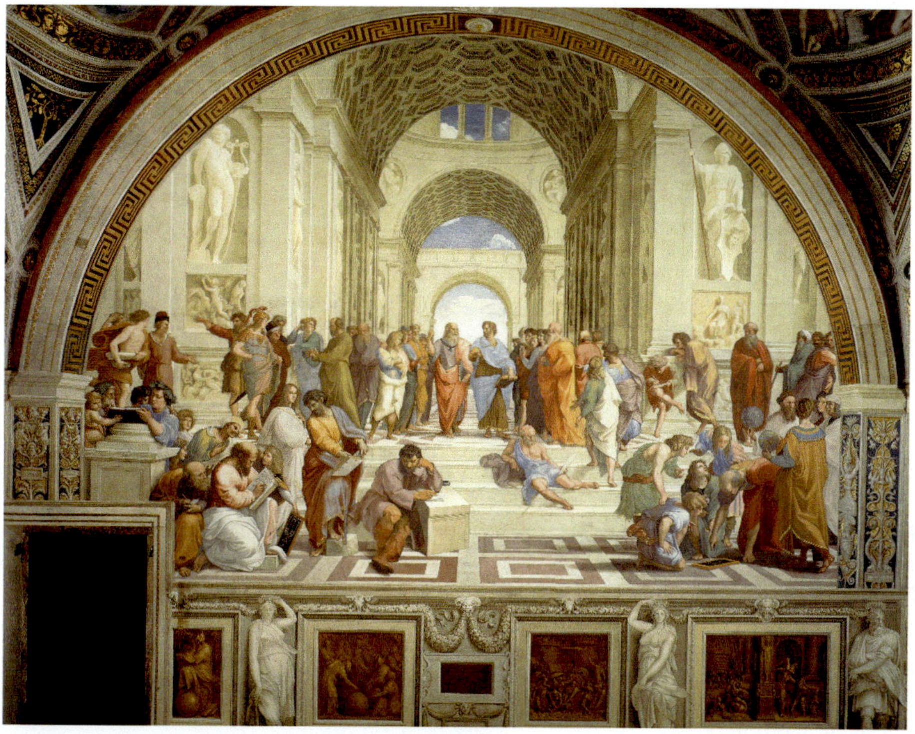

**Fig. 3.3** *School of Athens* (**Raphael**)—Stitched together from vatican.va, Public Domain, https://commons.wikimedia.org/w/index.php?curid=4406048. c. 1511; Raphael Rooms, Apostolic Palace, Vatican City

Now compare them with the Renaissance works in Figs. 3.3, 3.4 and 3.5.

The paintings in Figs. 3.1 and 3.2 are flat, two dimensional. There is no sense of depth in the painting; they are unrealistic. Look at how poorly St. Francis's (stigmatic) feet are painted; people's feet don't bend like that. But the paintings in Figs. 3.3, 3.4 and 3.5 are realistic, they draw you in, they seem to be three dimensional, or at least to convey a third dimension. They have *perspective*. What is perspective, and how does an artist achieve it?

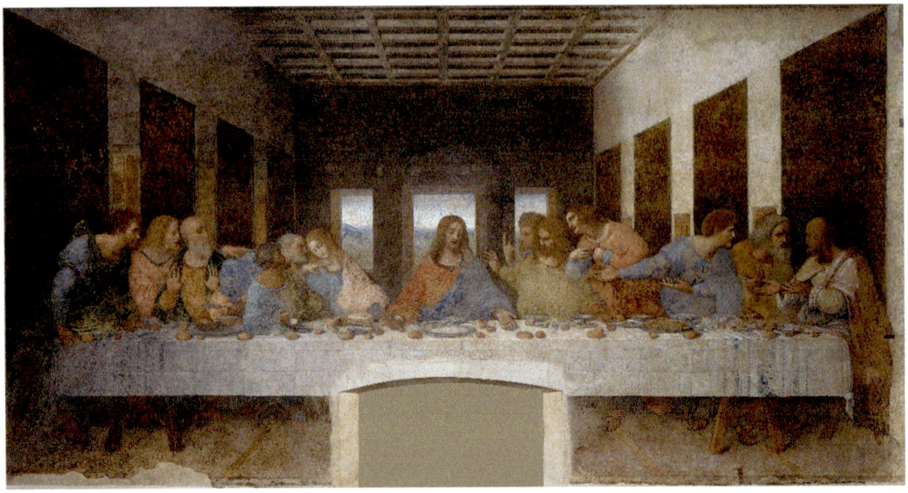

**Fig. 3.4** *The Last Supper* (**da Vinci**)—Unknown source, Public Domain, https://commons.wikimedia.org/w/index.php?curid=24759. c. 1498; Dining Hall Wall in Santa Maria delle Grazie Convent, Milan

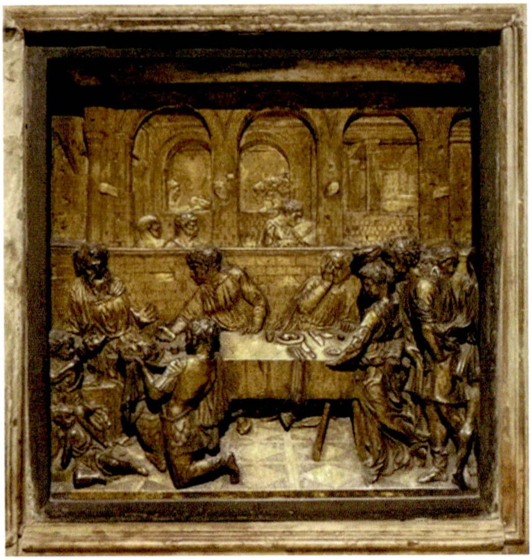

**Fig. 3.5** *Feast of Herod* (**Donatello**) By Donatello—Own work, CC BY 3.0, https://commons.wikimedia.org/w/index.php?curid=44222907. 1420; Battistero di San Giovanni, Sienna

## 3.2 Quantifying Perspective

We propose that a painting of a scene *has perspective* if the image it presents to the eye is identical to what one would see if the actual scene were viewed through a frame of glass. If one could paint, on a glass window, exactly what one sees through the window, then the painting would have perspective. I tried my hand at this in Fig. 3.6; one image is a photograph of a landscape and an empty frame, and the other is the same landscape with the frame containing my painting. Can you tell which is which? (The answer will be revealed at the end of this section.)[2]

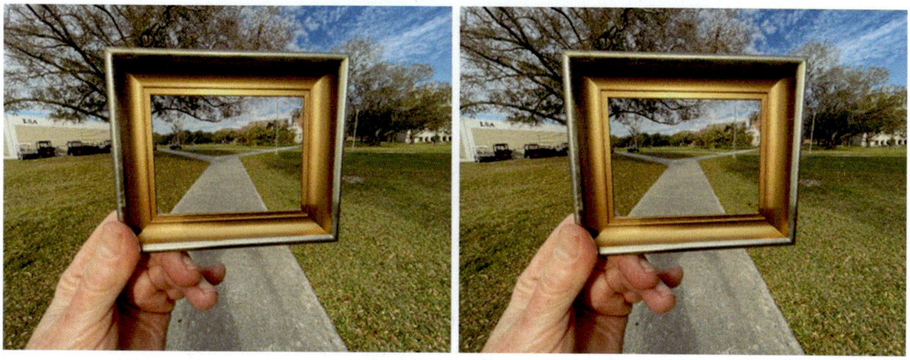

**Fig. 3.6** Perfect perspective

[2] **Rene Magritte**'s renditions of *The Human Condition* illustrate this notion:
https://en.wikipedia.org/wiki/The_Human_Condition_(Magritte).
https://artuk.org/discover/stories/rene-magrittes-la-condition-humaine.
https://www.clevelandart.org/art/1992.274.
https://www.wikiart.org/en/rene-magritte/the-human-condition-1935.

In fact Albrecht Durer took this quite literally with three of his woodcuts (Figs. 3.7, 3.8 and 3.9):

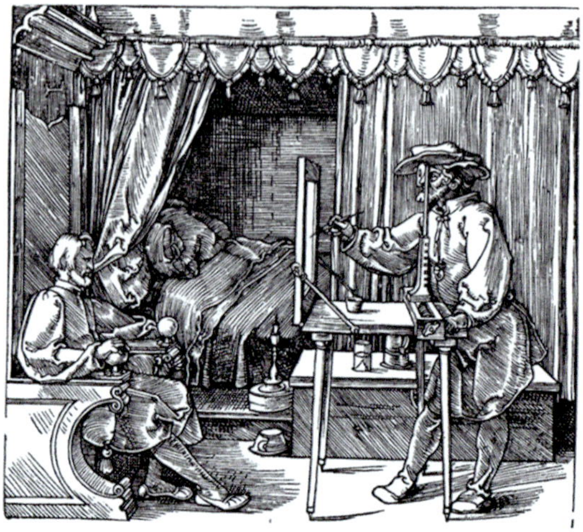

**Fig. 3.7** *Artist Drawing a Seated Man* (**Albrecht Durer**)—http://www.all-art.org/baroque/portrait1.html, Public Domain {{PD-US}}, https://commons.wikimedia.org/w/index.php?curid=19351437. 1525

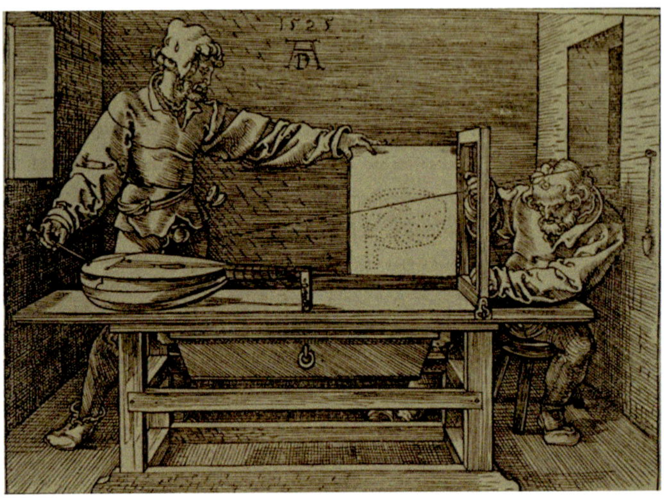

**Fig. 3.8** *Man Drawing a Lute* (**Dürer**)—http://www.usc.edu/schools/annenberg/asc/projects/comm544/library/images/626.html, Public Domain, https://commons.wikimedia.org/w/index.php?curid=1206785. 1525; Metropolitan Museum of Art, New York

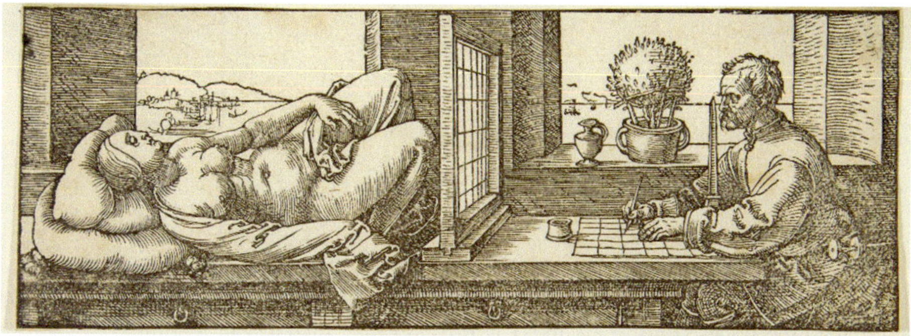

**Fig. 3.9** *Draughtsman Making a Perspective Drawing of a Reclining Woman* (***Dürer***)—This file was donated to Wikimedia Commons as part of a project by the Metropolitan Museum of Art. See the Image and Data Resources Open Access Policy, CC0, https://commons.wikimedia.org/w/index.php?curid=60859065. c. 1500; Metropolitan Museum of Art, New York

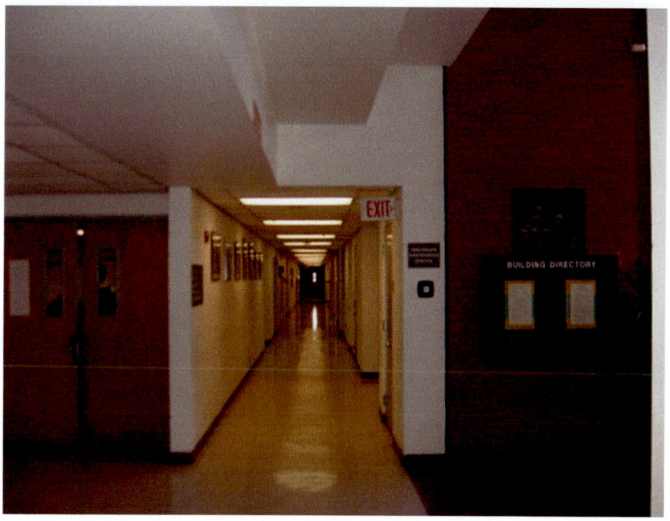


**Fig. 3.10** Straight ahead

Of course we don't want to use Durer's contraptions. To discern the mathematical principle behind perspective, try out this experiment in a long hallway (Figs. 3.10 and 3.11). Standing in the middle of the hallway, set up a large glass frame on an easel. Now there are four parallel lines running the length of the hall, where the floor intersects the walls and where the walls intersect the ceiling. Using an erasable marker, trace these lines on the glass; do not shift your head as you do so. Now use a straightedge to extend the lines. You will find they intersect in a single point.

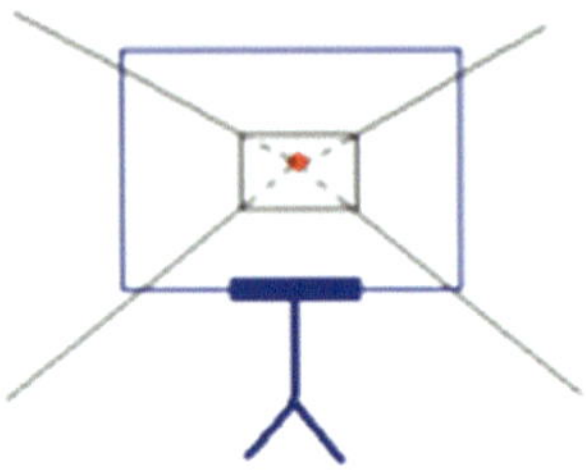

**Fig. 3.11** Parallel lines in Fig. 3.10

Now shift your easel to one side, still looking down the hallway. Trace the four lines and extend the traces (Figs. 3.12 and 3.13). Again they intersect in a point.

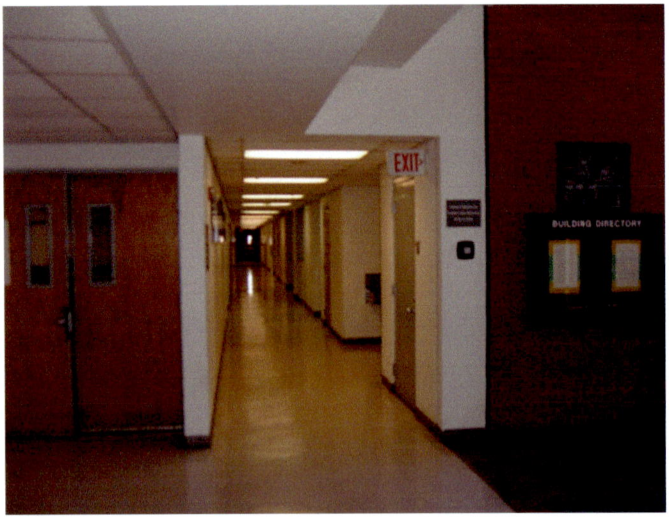


**Fig. 3.12** Slight angle

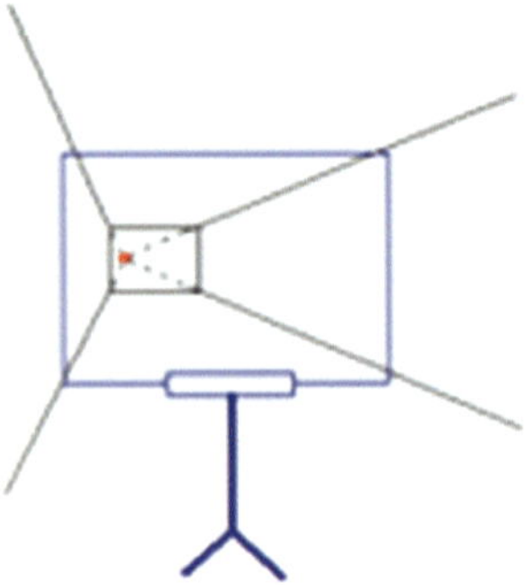

**Fig. 3.13** Parallel lines in Fig. 3.12

This time stand in the middle but turn at an angle to the hallway, so that only two of the lines are visible (Figs. 3.14 and 3.15). Trace the lines and extend. Once again, they intersect in a point. The point may be off the glass, but we could recover it by using a larger pane of glass.

The main principle of perspective is that *any group of lines that are parallel in 3-D space* will *intersect in a* ***single*** *point* in the plane of the painting. That point depends on the set of parallel lines, and may lie beyond the edge of the painting. (To be precise, if the lines happen to be parallel to the plane of the painting in 3-D, they will be parallel in

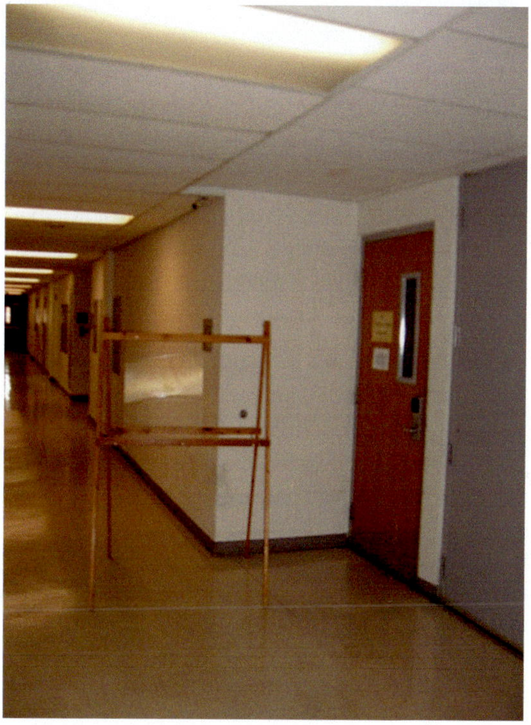

**Fig. 3.14** Extreme angle

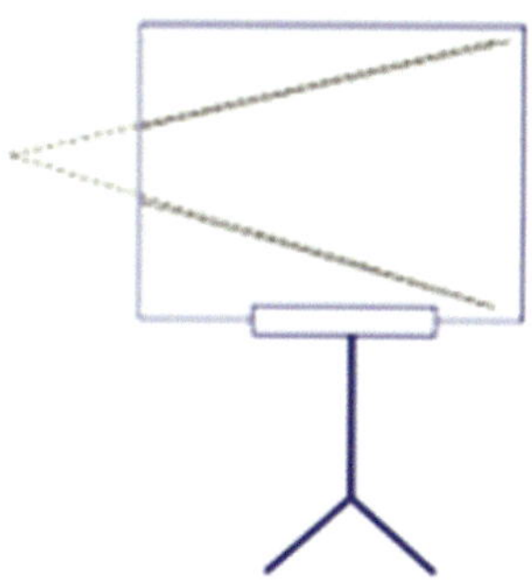

**Fig. 3.15** Parallel lines in Fig. 3.14

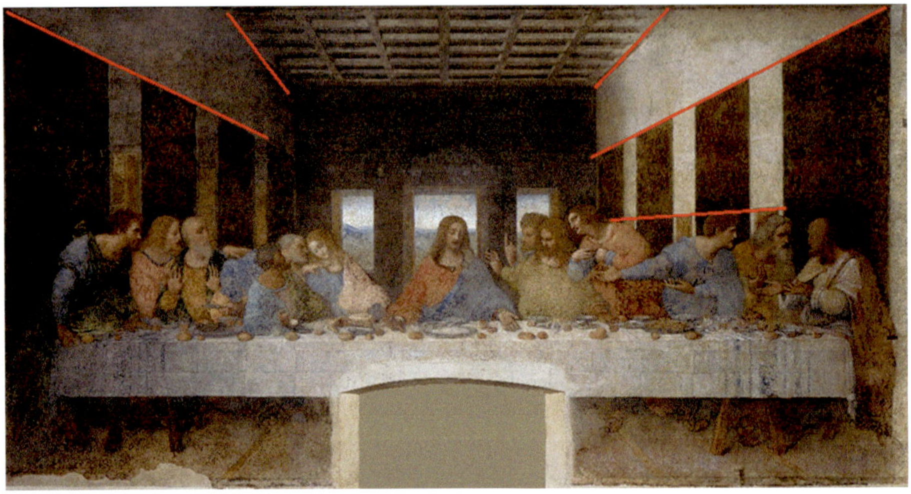

**Fig. 3.16** *The Last Supper* (**da Vinci**)—Unknown source, Public Domain, https://commons.wikimedia.org/w/index.php?curid=24759. c. 1498; Dining Hall Wall in Santa Maria delle Grazie Convent, Milan (Enhanced from original)

the painting also and will not intersect.) Artists call the point of intersection the *vanishing point* for the lines. *Sometimes "THE" vanishing point refers to the (single) vanishing point peculiar to the set of lines running perpendicular to the painting.*

It is amusing to trace these lines for Fig. 2.87, The Last Supper: guess where they intersect? Download the image and use your drawing tool to extend the parallel lines (Figs. 3.16 and 3.17).

Surprised?

Did Leonardo really use this procedure to create his paintings? Yes he did! Look at this preliminary study (Fig. 3.18).

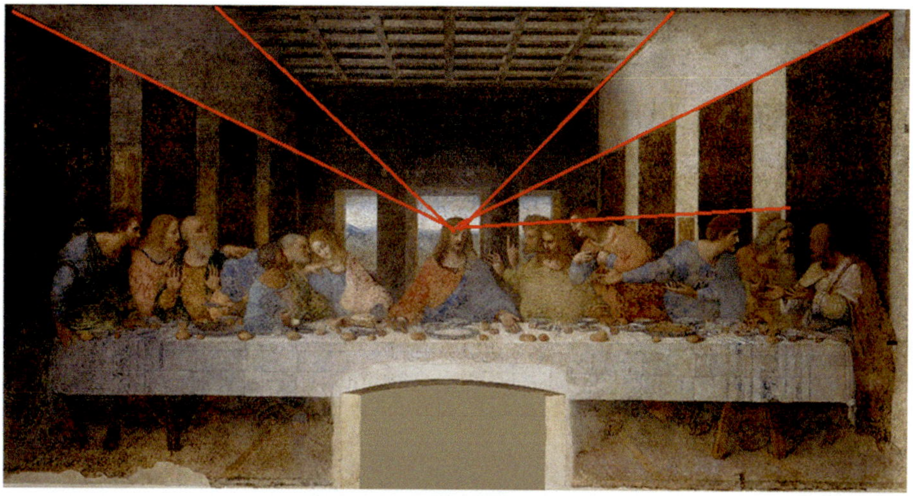

**Fig. 3.17** *The Last Supper* (**da Vinci**)—Unknown source, Public Domain, https://commons.wikimedia.org/w/index.php?curid=24759. c. 1498; Dining Hall Wall in Santa Maria delle Grazie Convent, Milan (enhanced from original)

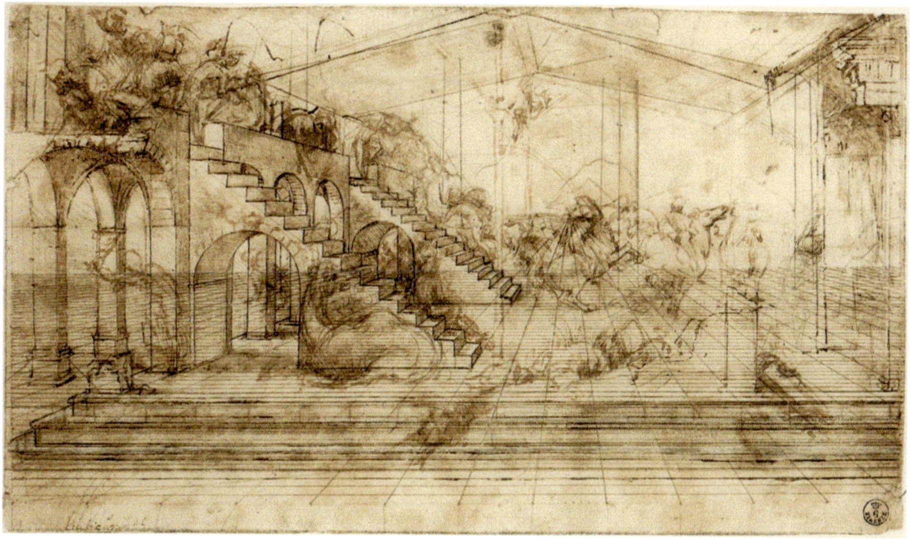

**Fig. 3.18** Study for *Adoration of the Magi* (**da Vinci**)—Bortolon, The Life and Times of Leonardo, Paul Hamlyn, Public Domain, https://commons.wikimedia.org/w/index.php?curid=1647260

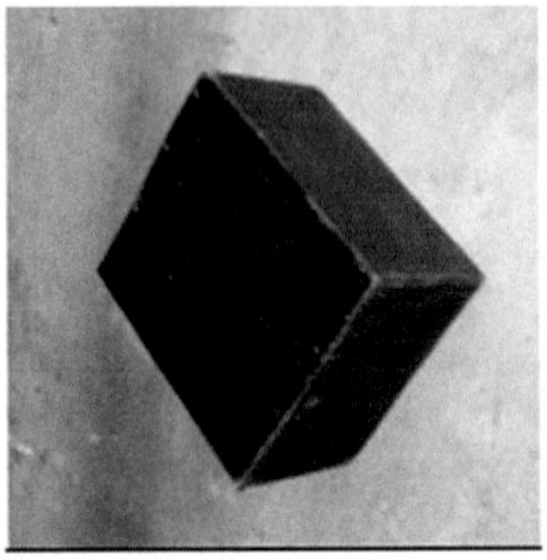

**Fig. 3.19** Vanishing point practice #2

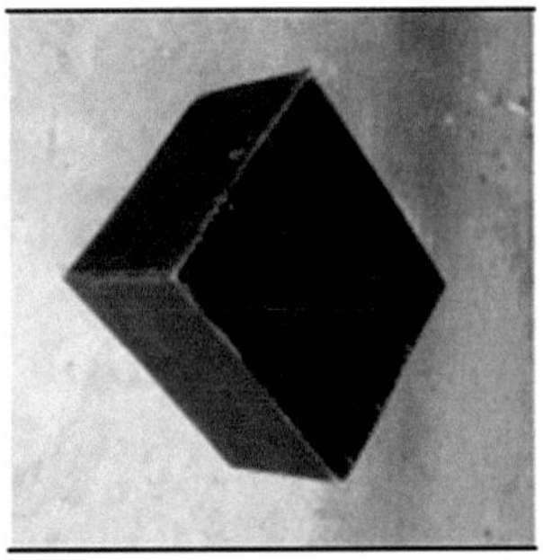

**Fig. 3.20** Vanishing point practice #2

Try this out yourself. Download *School of Athens* and *Feast of Herod* (Figs. 3.3 and 3.5) and use the straight line tool in your drawing app to find the vanishing points.

EXERCISE. Find *all* of the vanishing points for the visible parallel edges of the boxes in Figs. 3.19 and 3.20.

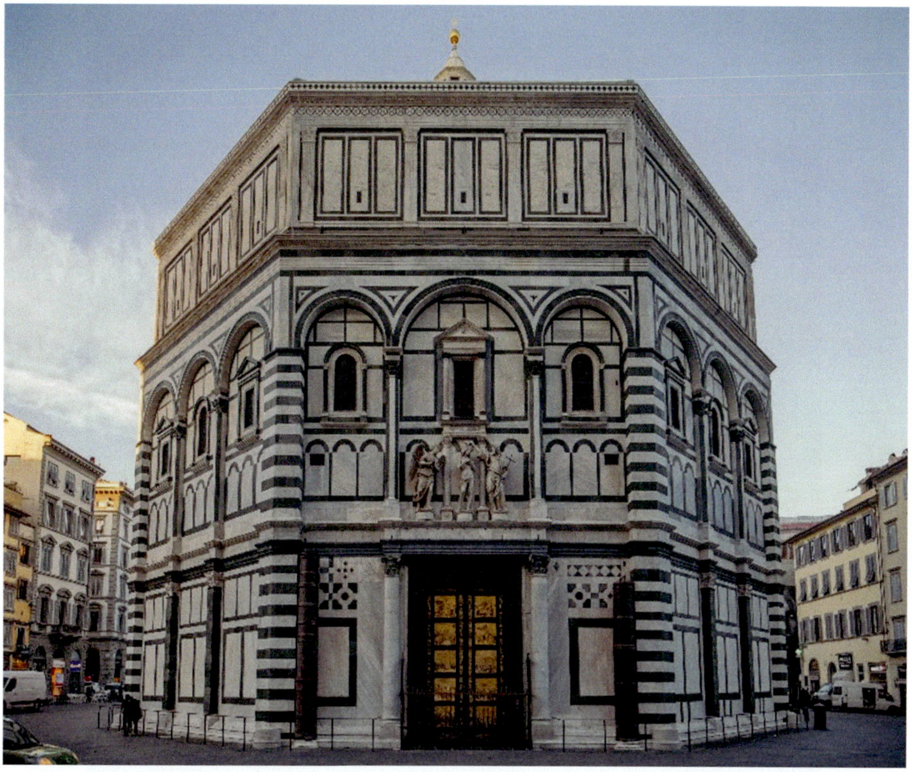

**Fig. 3.21** Baptistry of St. John, Florence By Bradley Weber—https://www.flickr.com/photos/41294655@N00/44920511001/, CC BY 2.0, https://commons.wikimedia.org/w/index.php?curid=100735889

To really test your mettle, see how many vanishing points you can find for these images (Figs. 3.21, 3.22 and 3.23).

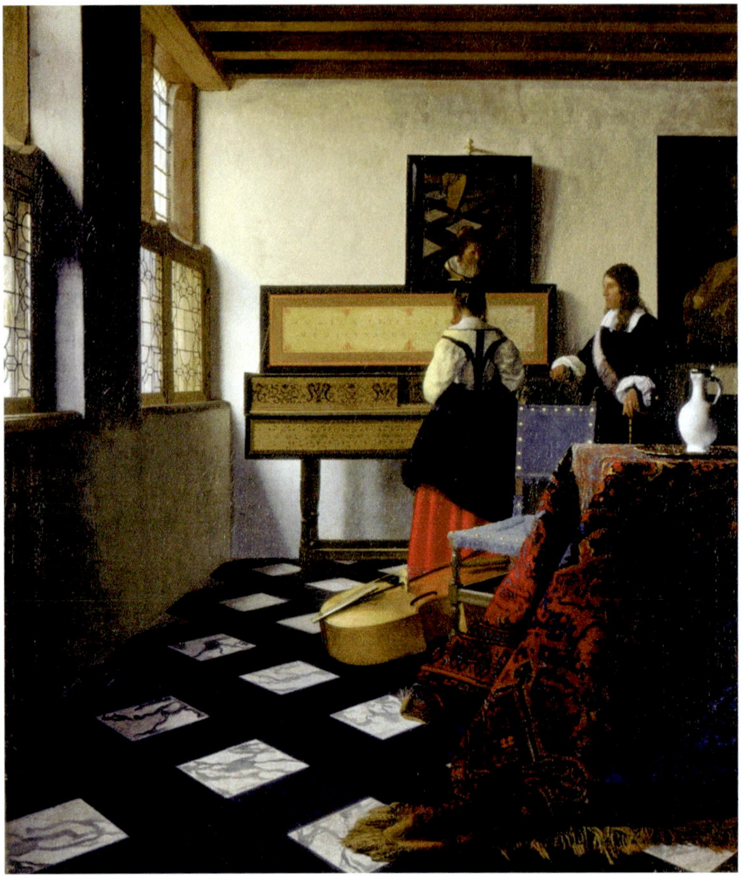

**Fig. 3.22** *The Music Lesson* (**Vermeer**)—CwFxCw3kUr-Lrw—Google Arts & Culture, Public Domain, https://commons.wikimedia.org/w/index.php?curid=22127311. 1662; Queen's Gallery at Buckingham Palace, London

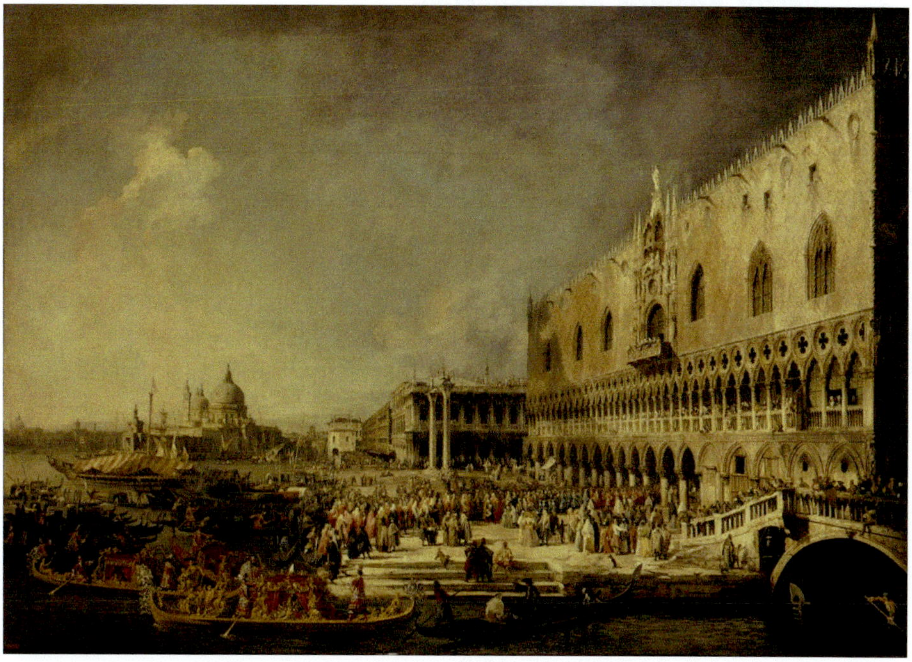

**Fig. 3.23** *Arrival of the French Ambassador in Venice* (**Canaletto**)—https://www.hermitagemuseum.org/wps/portal/hermitage/digital-collection/01.%20Paintings/32231?lng=en, Public Domain, https://commons.wikimedia.org/w/index.php?curid=126593508. c. 1927; Hermitage Museum, Saint Petersburg

ANSWER TO THE Fig. 3.6 QUESTION.

You STEMs are soooo easy. Both photographs are the same. I couldn't paint like that if my life depended on it.

I can barely draw the partial derivative operator.

## 3.3 Vector Analysis of Perspective

Why do all lines that are parallel in three dimensions intersect in a single point in a 2D rendering? If you're an engineer, now's your chance to strut your stuff for your art friends. You know vector calculus!

**Mathematical Analysis** A line in space generates a line on the canvas whose points are given by the intersections of the canvas and the lines of sight from the viewer's eye to the points on the space-line (Figs. 3.7, 3.8, 3.9, 3.10, 3.11, 3.12, 3.13, 3.14 and 3.15). As the points on two parallel lines in space become more remote ("approach infinity"), the lines of sight approach a limiting line: namely, the line from the eye that is parallel to the space line (Fig. 3.24).

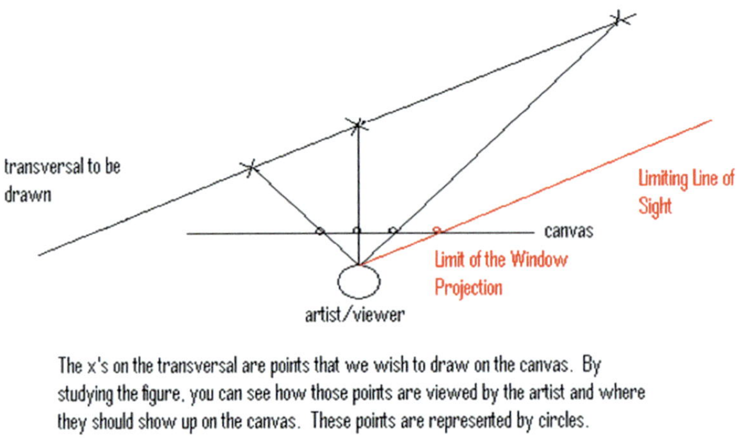


**Fig. 3.24** Limit of line of sight

In vector language, let $\mathbf{R}_0$ be a point on the space line, and let $\mathbf{v}$ be a unit vector parallel to the space line. Then the space line is given parametrically as

$$\mathbf{R}(t) = \mathbf{R}_0 + t\mathbf{v}, \quad -\infty < t < \infty.$$

If $\mathbf{R}_{eye}$ is the vector locating the eye, then the line of sight is given by

$$\mathbf{R}_{LOS}(t) = \mathbf{R}(t) - \mathbf{R}_{eye} = \mathbf{R}_0 + t\mathbf{v} - \mathbf{R}_{eye},$$

and its limiting direction is $\mathbf{v}$:

$$\lim_{t\to\infty} \mathbf{R}_{LOS}(t)/t = \lim\{\mathbf{R}_0/t + \mathbf{v} - \mathbf{R}_{eye}/t\} = \mathbf{v}.$$

Therefore, on the canvas, the images of the points of the space line converge to the point where the line from the eye, parallel to $\mathbf{v}$, strikes the canvas (or the canvas, extended).

The surprise is that this point is independent of $\mathbf{R}_0$, so *all* space lines *parallel* to $\mathbf{v}$ generate canvas images that converge to this same point—the "vanishing point" for space lines in this direction (Fig. 3.25).

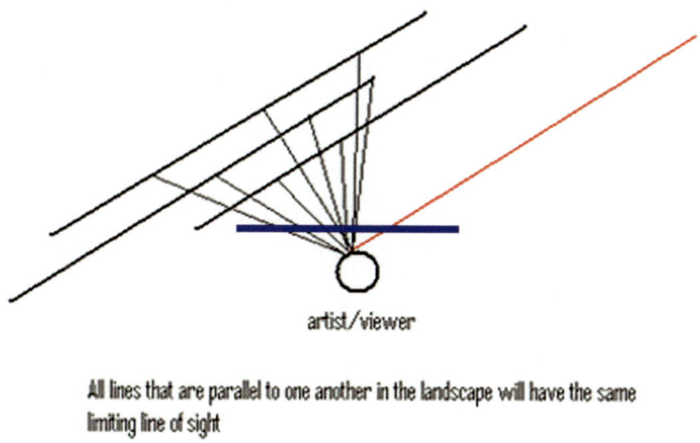


**Fig. 3.25** The vanishing point

Note that accurate perspective depends on the location of the eye. It usually doesn't make much difference, but the vanishing point perspective in the right eye is not exactly the same as that in the left. The artists in the first and third Durer woodcuts (Figs. 3.7 and 3.9) are shielding one eye.

And to reiterate:

(a) the vanishing point may lie beyond the canvas (Fig. 3.26);
(b) if the original lines are parallel to the canvas, the vanishing point*s* are "at infinity," in opposite directions (Fig. 3.27).

In fact if the original line is vertical, one vanishing point is directly above (the "zenith") and the other is directly below (the "nadir") the viewer.

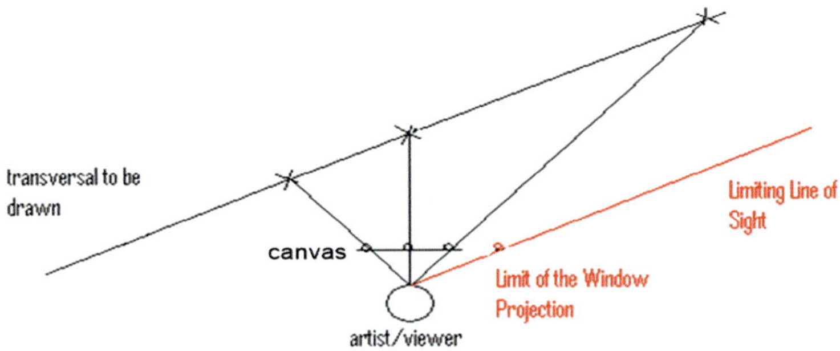


**Fig. 3.26** Vanishing point lies off the canvas

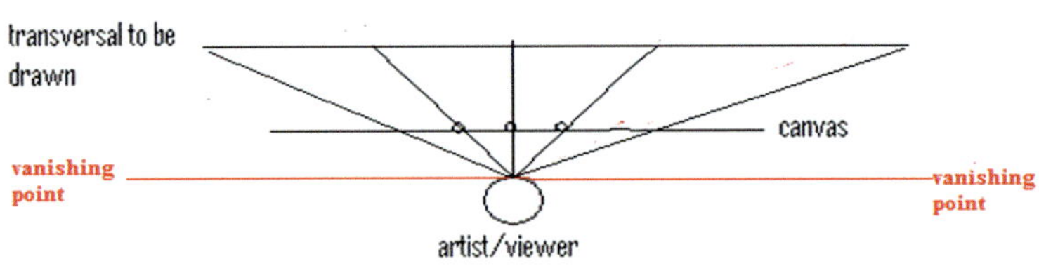


**Fig. 3.27** Vanishing points at infinity

## 3.4 The Discovery of Perspective

Who discovered the vanishing point principle? History credits Filippo Brunelleschi (1377–1446), an engineer from (where else?) Florence. The discovery came about as a result of a competition. Across the plaza from the Duomo cathedral in Florence, there is the Baptistery of St. John (Fig. 3.28) (If you search the web for "Virtual Tours of Florence" you can find several sites that provide excellent images of the Duomo and the Baptistry - exterior and interior).

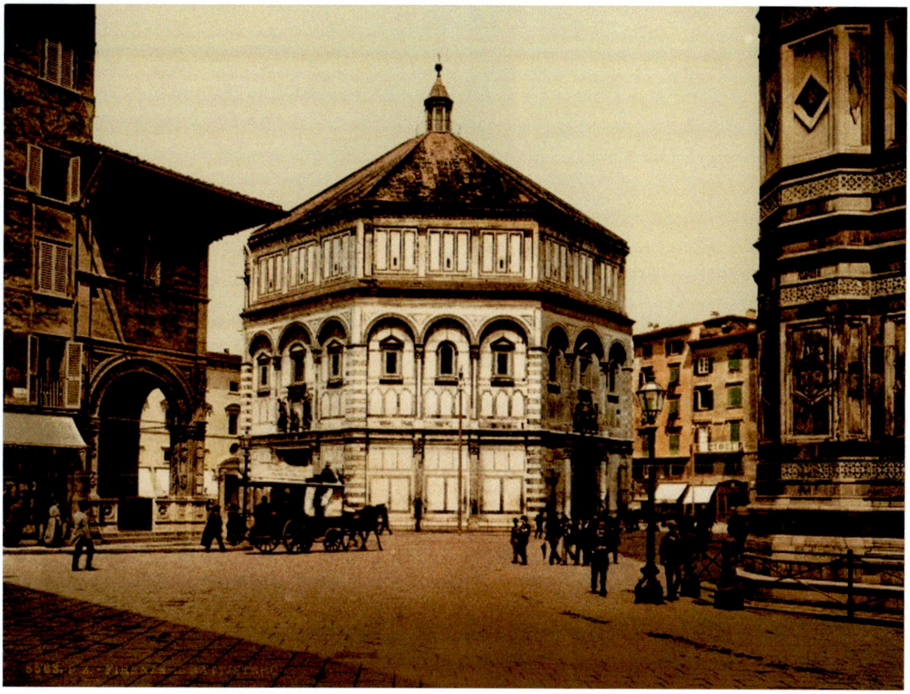

**Fig. 3.28** Baptistry of St. John, Florence—By Photoglob Zürich, between 1890 and 1900. From the Photochrom Prints Collection at the Library of Congress. Public Domain. https://www.flickr.com/photos/trialsanderrors/4846282902

The Baptistry is octagonal. This means that it is hellish to draw realistically (Figs. 3.29 and 3.30).[3]

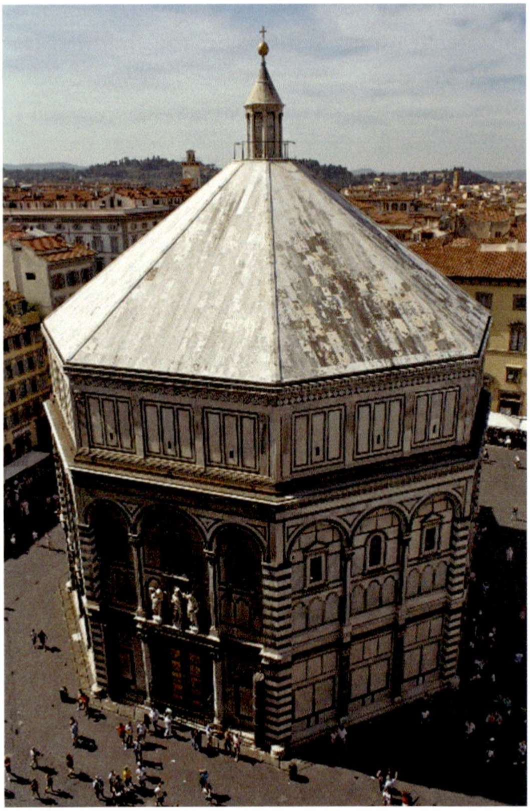

**Fig. 3.29** Florence Baptistery—Photo by Alexander Baxevanis. CC BY 2.0 https://www.flickr.com/photos/futureshape/6212060793/

[3] Other hapless medieval attempts predating the mastery of perspective can be viewed at

https://commons.wikimedia.org/wiki/File:Totila_fa_dstruggere_la_citt%C3%A0_di_Firenze.jpg.

https://www.academia.edu/13019977/Representation_of_Florentine_History_and_Creation_of_Communal_Myths_in_the_Illustrated_Nuova_Cronica_of_Giovanni_Villani.

www.artandarchitecture.org.uk/fourpaintings/daddi/popup.html?image=large/d19.jpg&title=TheMadonnaofMercy.

https://www.pinterest.co.uk/pin/644859240366382388/.

https://www.pinterest.com/pin/152911349835338857/.

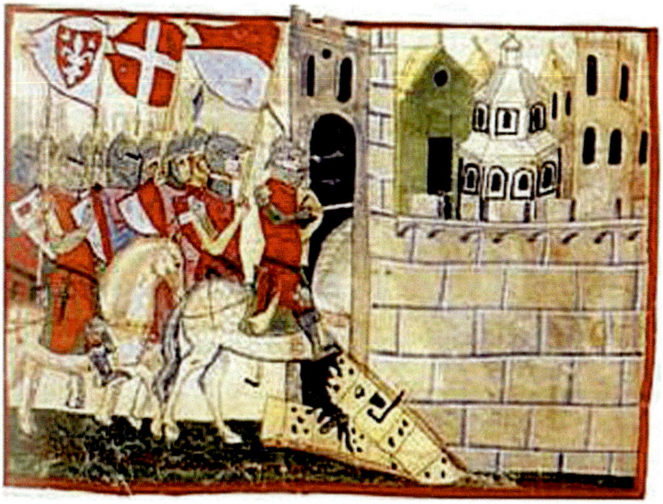

**Fig. 3.30** Illustration from Giovanni Villani's *Cronica Nuova* (14th c.) Public Domain http://upload.wikimedia.org/wikipedia/commons/a/af/Nuova2Florence.jpg; https://manuscriptminiatures.com/image/8561 (cropped from original)

The city of Florence sponsored a contest in the early 1400s to see who could draw the Baptistry in perspective—*and prove* that their painting exactly matched what the eye saw. Brunelleschi won the contest by drilling a small hole in a mirror and placing the mirror on his easel so that it reflected the Baptistry. He then traced the reflected image directly onto the surface of this mirror and painted the trace, giving him a faithful reproduction of the Baptistry on his (pierced) mirror "canvas" (Fig. 3.31).

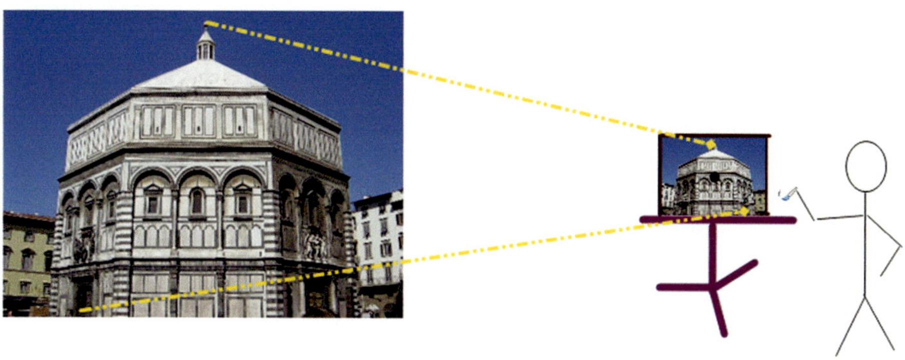

**Fig. 3.31** Mirror image of Baptistery by Christopher Kaetz—Self-photographed, Public Domain, https://commons.wikimedia.org/w/index.php?curid=4518767 (cropped and embellished from the original)

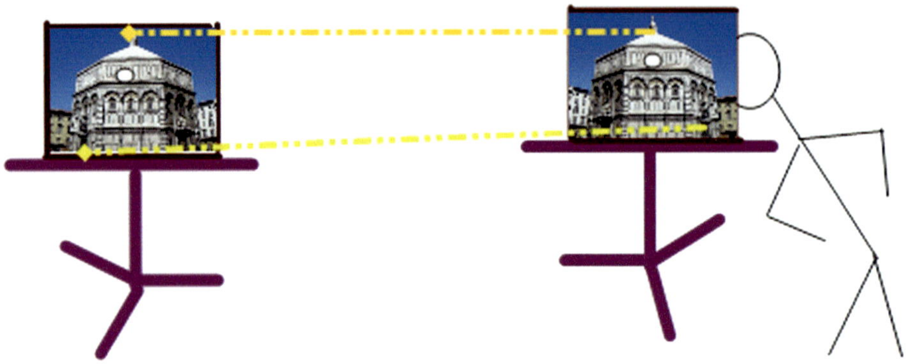

**Fig. 3.32** Mirror Image of Mirror Image of Baptistry by Christopher Kaetz—Self-photographed, Public Domain, https://commons.wikimedia.org/w/index.php?curid=4518767 (cropped and embellished from the original)

Well, to be honest, it's a faithful reproduction of the mirror *reflection* of the Baptistry; it has to be "re-reflected" to replicate the Baptistry exactly. Here's where the hole comes in. By looking through the hole, *from the back of the mirror*, at a *second* mirror, one would see a reflection of Brunelleschi's painting—that is, an exact, correctly oriented, image of the Baptistry (Fig. 3.32).

The historical literature is a little inconsistent here, but Brunelleschi probably traced and painted the image on the second mirror to achieve the meticulous portraiture prescribed by the contest judges. Its authenticity could be established by moving it in and out of one's line of view of the Baptistry itself (Fig. 3.33).

The first site below has a 9 min URL explaining Brunelleschi's process. The others contain various images to help you get the idea.

https://www.youtube.com/watch?v=frB-XiXeNiQ

https://maitaly.files.wordpress.com/2011/04/0328p_duomo6_b.jpg

https://www.edgeofyesterday.com/time-travelers/filippo-brunelleschi-sculptor-goldsmith-duomo-maker

https://en.wikipedia.org/wiki/Filippo_Brunelleschi#/media/File:Brunelleschi's_perspective_experiment.jpg

https://www.researchgate.net/publication/312659427_Depth_Perception_and_the_History_of_Three-Dimensional_Art_Who_Produced_the_First_Stereoscopic_Images/Fig.s?lo=1

http://www.manthano.fr/pages/La%20Renaissance.html

https://ljojo.com/blog/2016/7/24/my-illustration-of-filippo-brunelleschis-linear-perspective-experiment-for-human-factors-in-simulation-and-training-book

By experimenting with his "sanctioned" perspective painting, Brunelleschi discerned the mathematical underpinnings of the vanishing point principle (Fig. 3.34).

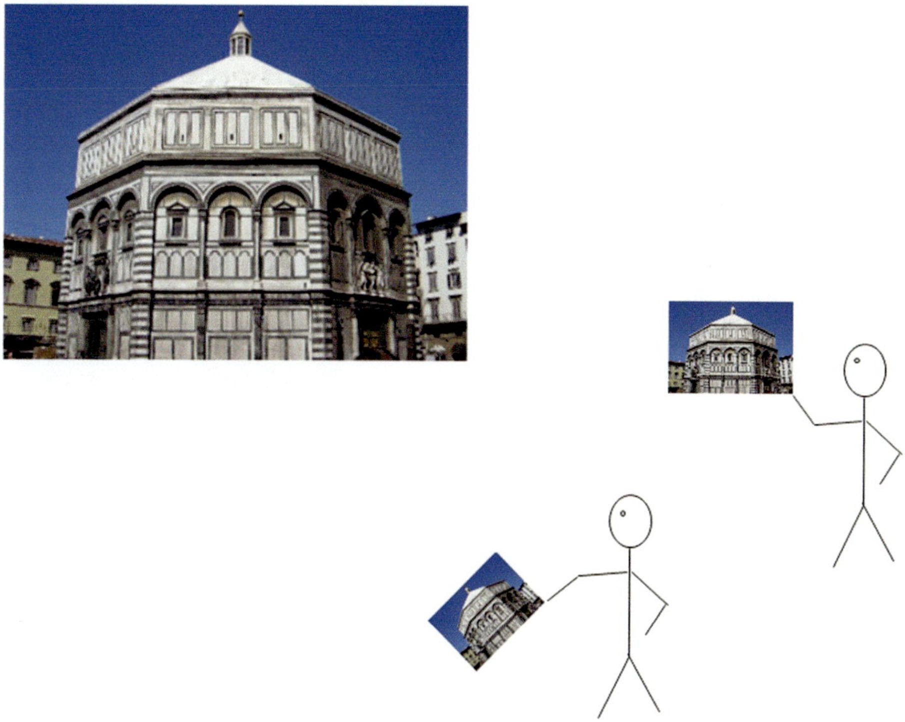

**Fig. 3.33** Verifying the authenticity of Brunelleschi's painting by Christopher Kaetz—Self-photographed, Public Domain, https://commons.wikimedia.org/w/index.php?curid=4518767 (cropped and embellished from the original)

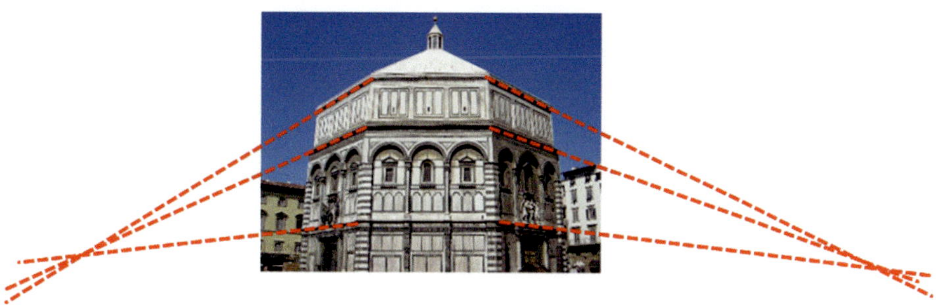

**Fig. 3.34** Vanishing Points of the Baptistery by Christopher Kaetz—Self-photographed, Public Domain, https://commons.wikimedia.org/w/index.php?curid=4518767 (cropped and embellished from the original)

In the context of perspective, many tourists have found this painting amusing (Fig. 3.35).

It is on an inside wall of the Florentine Duomo cathedral. The perspective is inconsistent. The top of the painting presumes the viewer is above the statue, but the bottom puts the viewer at its base. (It may help if you zoom in on the picture.) And the horse has both left legs off the ground!

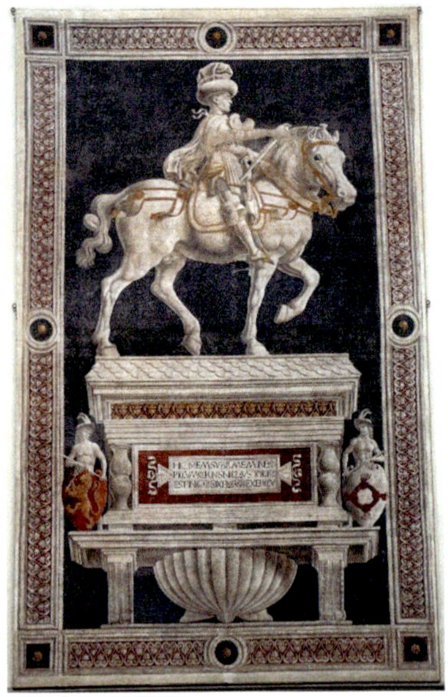

**Fig. 3.35** *Equestrian Monument of Niccolò da Tolentino*, 1456 (**Andrea del Castagno**)—By Sailko, CC BY 3.0 <https://creativecommons.org/licenses/by/3.0>, via Wikimedia Commons; https://commons.wikimedia.org/wiki/File:Andrea_del_castagno,_Monumento_equestre_di_Niccol%C3%B2_da_Tolentino,_1456,_01.JPG

## 3.5 Filippo Brunelleschi

Brunelleschi (Fig. 3.36) initially was a goldsmith. Sometime before discovering the perspective principle, he had entered another competition sponsored by the city of Florence in 1401: to decorate the north doors for the Florence Baptistry. The building had an older set of doors, designed in 1336 by Andrea Pisano et al. and moved from the east wall to the south wall in 1452. Twenty of the south door panels depict scenes from the life of John the Baptist, and the others depict the eight virtues. Closeups can be viewed at https://www.wga.hu/html_m/p/pisano1/andrea/1southdo/index.html. The panel in Fig. 3.37 displays the Angel of Hope (Close inspection reveals he has a hexagonal halo).

Lorenzo Ghiberti required 22 years to create the North Doors (Fig. 3.38). Twenty of the panels depict the life of Christ, and eight show evangelists and Church Fathers. They can be viewed at https://www.wga.hu/html_m/g/ghiberti/north_do/index.html.

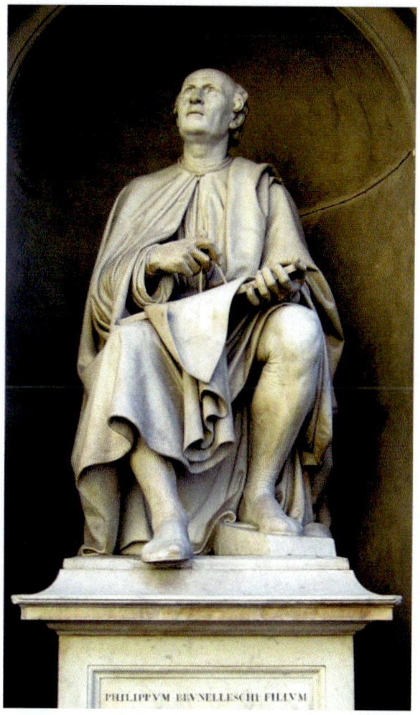


**Fig. 3.36** *Brunelleschi* **(Luigi Pampaloni)** No machine-readable author provided. Richardfabi~commonswiki assumed (based on copyright claims), Public Domain, via Wikimedia Commons; https://commons.wikimedia.org/wiki/File:Bunelleschi.jpg 1838; Piazza del Duomo, Florence

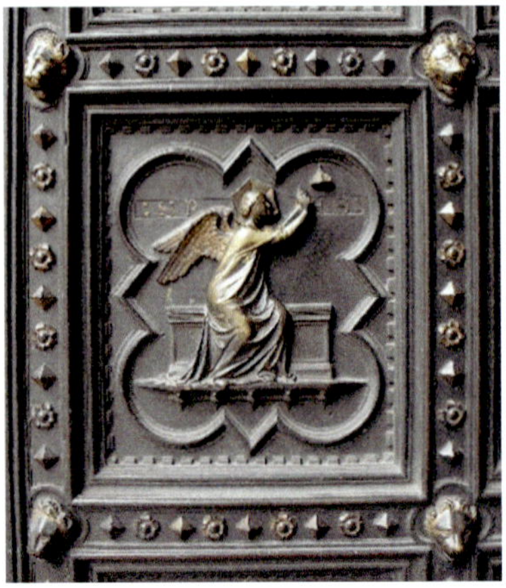

**Fig. 3.37** *Hope* Baptistery South Doors detail (**Pisano**)—Photo by Helena https://www.flickr.com/photos/heleenmeijer/6831636709 CC BY 2.0 Deed. 1330; Florence

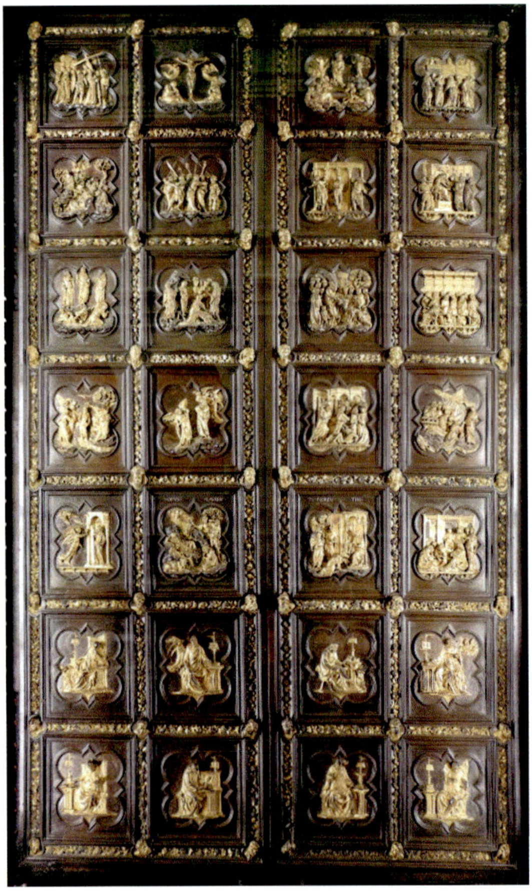

**Fig. 3.38** Baptistery North Doors (**Ghiberti**) By Sailko—Own work, CC BY 3.0, https://commons.wikimedia.org/w/index.php?curid=45833936. 1402–24; Florence

The North Doors earned Ghiberti acclaim as an accomplished artist and in 1425 he received a second commission, to decorate the east doors.[4] This proved to be his masterpiece. The work comprises ten panels depicting stories from the Old Testament. Incorporating the principles of perspective formulated by Brunelleschi, Ghiberti's East Doors were called the *Gates of Paradise* by Michelangelo (Fig. 3.39, 3.40, 3.41, 3.42 and 3.43).

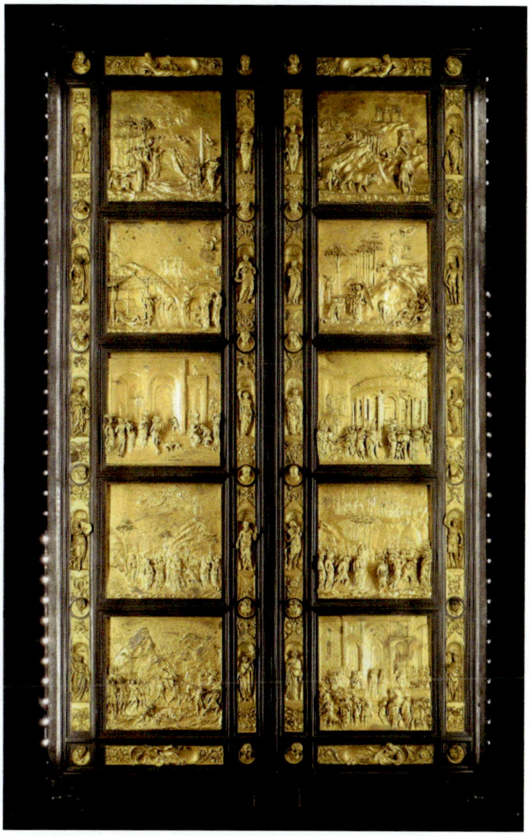

**Fig. 3.39** Baptistry East Doors *Gates of Paradise* (**Ghiberti**)—By Sailko—Own work, CC BY 3.0, https://commons.wikimedia.org/w/index.php?curid=45878184. 1425–52; Florence

[4] (The west wall contains an apse).

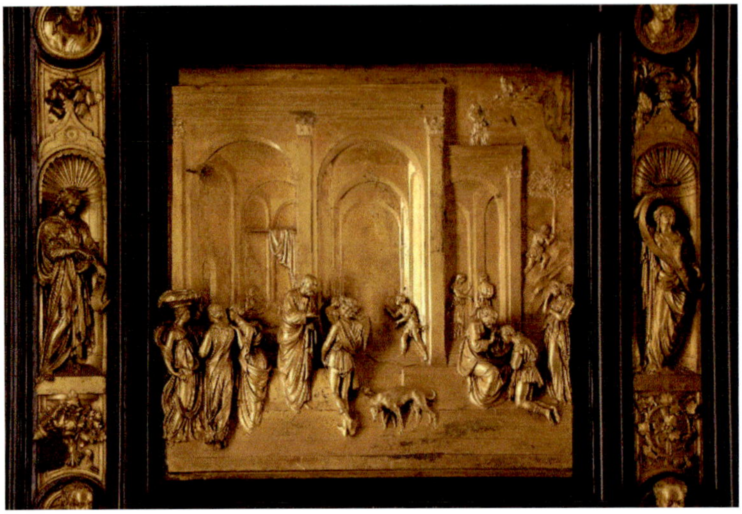

**Fig. 3.40** *Jacob and Esau* Florence Baptistery detail (**Ghiberti**)—By Sailko—Own work, CC BY 3.0, https://commons.wikimedia.org/w/index.php?curid=45878456. 1425–52; Florence

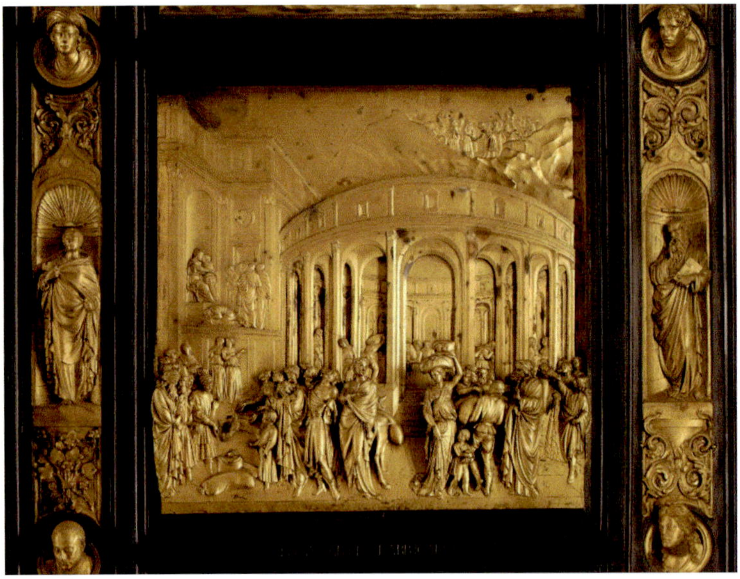

**Fig. 3.41** *Joseph and His Brothers* Florence Baptistery detail (**Ghiberti**)—By Sailko—Own work, CC BY 3.0, https://commons.wikimedia.org/w/index.php?curid=45878457. 1425–52; Florence

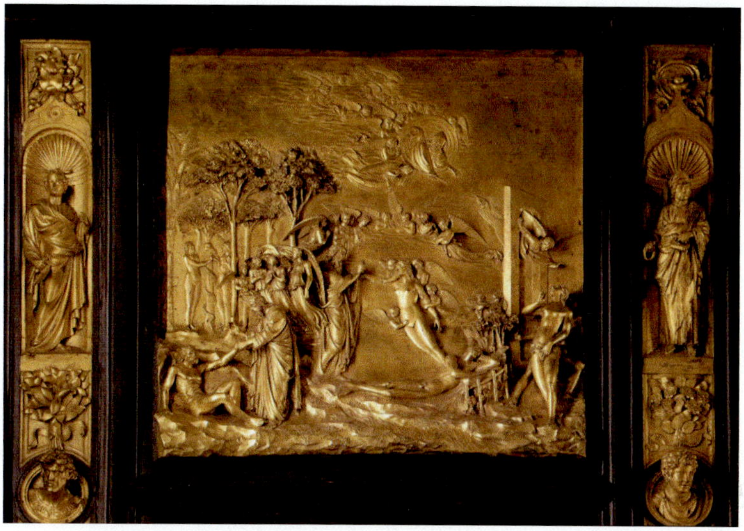

**Fig. 3.42** *Adam and Eve* Florence Baptistery detail (**Ghiberti**)—By Sailko—Own work, CC BY 3.0, https://commons.wikimedia.org/w/index.php?curid=45878455. 1425–52; Florence

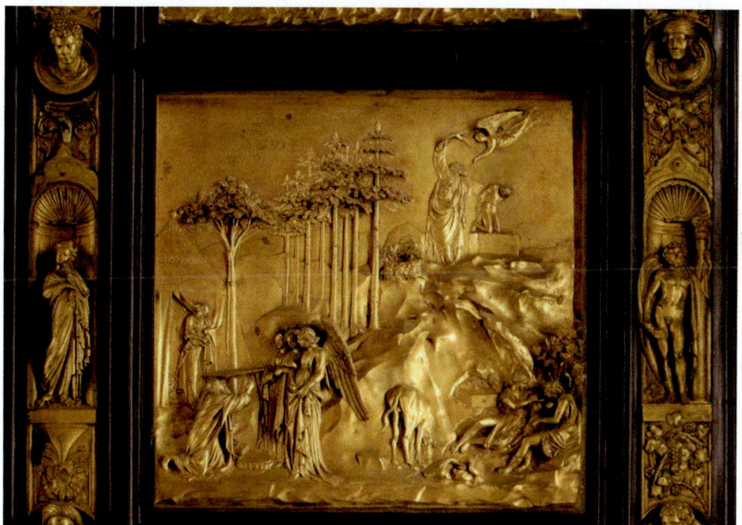

**Fig. 3.43** *Abraham* Florence Baptistery detail (**Ghiberti**)—By Sailko—Own work, CC BY 3.0, https://commons.wikimedia.org/w/index.php?curid=45878454. 1425–52; Florence

Photographs of all of the east door panels can be found at

https://homepages.bluffton.edu/~sullivanm/italy/florence/ghibertiparadise/ghibertiparadise.html

https://homepages.bluffton.edu/~sullivanm/italy/florence/duomomuseo/paradisedoors.html

https://homepages.bluffton.edu/~sullivanm/ghibertinorth/ghibertinorth.html

In 1943 the Nazi command in Italy ordered the evacuation of portable works of art in Florence, including the Gates of Paradise. The city hid the doors in a railroad tunnel until Florence's liberation in 1944. Bruno Bearzi, then caretaker of Florence's art works, allowed molds to be made of the panels. Replicas made from these molds can be seen in Grace Cathedral in San Francisco and in Trinity Lutheran Church in Hicksville, NY. You can also see replicas in the Victoria and Albert Museum in London, the Peabody Art Collection in Baltimore, the Getchell Library at the University of Nevada at Reno, the Forest Lawn Cemetery in Glendale, and the Harris Museum and Art Gallery in Preston, England.

After the Arno river flooded Florence in 1966, the panels were secured in the Museo dell'Opera del Duomo and a replica door, commissioned by Japanese businessman Choichiro Motoyama, was installed in the Baptistry in 1990. (Some of Figs. 3.40, 3.41, 3.42 and 3.43 are photographs of the replicas.)

Back to Brunelleschi. In 1418 he entered another contest (!) sponsored by the city to build the dome for its unfinished cathedral. Many artisans (Ghiberti among them) vied for this commission. Brunelleschi claimed that he could build the dome without using scaffolding, but he refused to submit a model or plan. Vasari's version of the story is re-told by Martin Gardner:

> The Italian architect Filippo Brunelleschi had designed a dome for a cathedral in Florence named the Santa Maria del Fiore. The city fathers demanded to see his model, but he refused. Instead, he challenged a group of architects to make an egg stand on end. Whoever succeeded, he told them, would be allowed to build the dome. After they all failed, he demonstrated how it could be done by tapping the egg on a marble table to flatten one end. The craftsmen protested that they could have done the same, but Filippo answered laughing that they could also have raised the cupola if they had seen his model. And so it was resolved that he should be commissioned to carry out this work. When the church was finally built … it had the shape of half an egg slightly flattened at the top.[5] (Martin Gardner, "The great egg-balancing mystery," *Skeptical Inquirer*, 20.3, May–June 1996, p. 8.)

[5] (This egg-balancing story was attributed to Christopher Columbus fifteen years later).

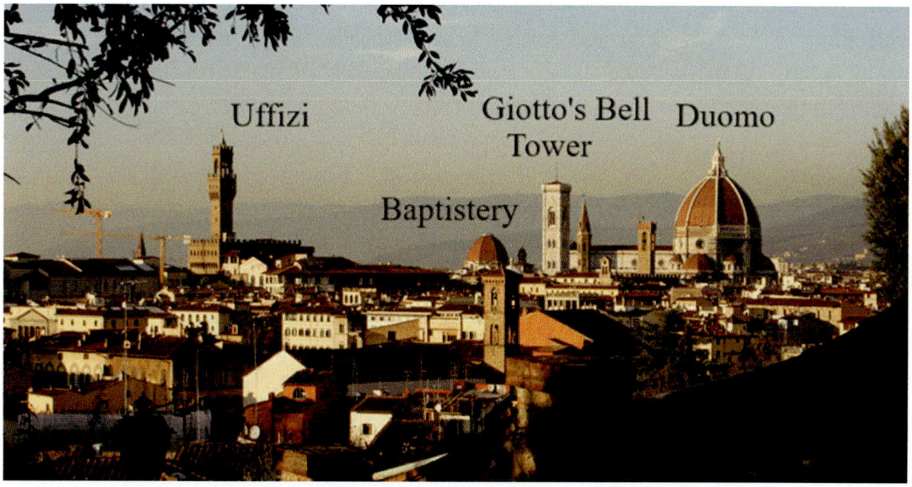


**Fig. 3.44** Florence Skyline—By Kari—CC BY 2.0 DEED https://www.flickr.com/photos/designsbykari/ Cropped and embellished from original

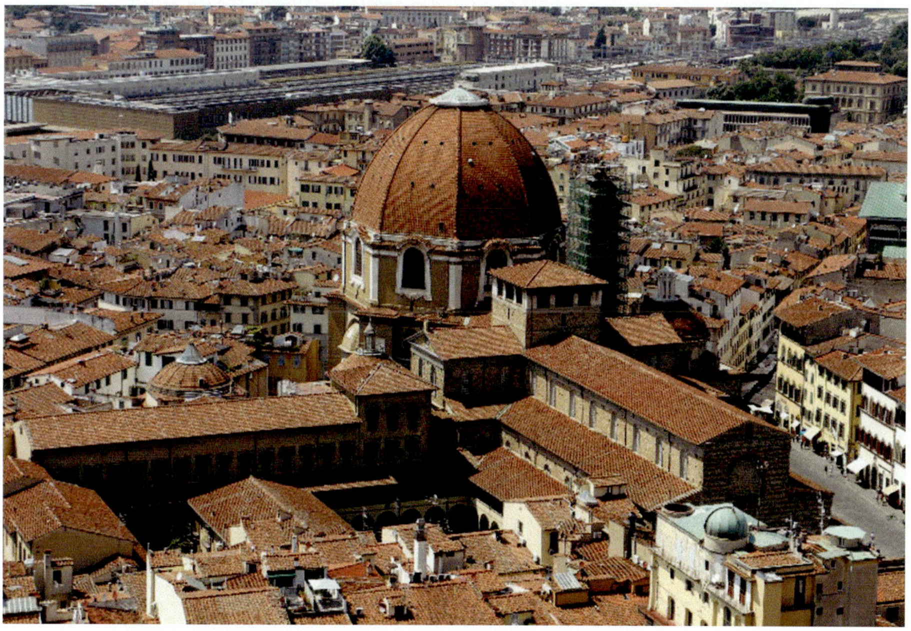

**Fig. 3.45** Florence Duomo Cupola—By Sailko—Own work, CC BY 3.0, https://commons.wikimedia.org/w/index.php?curid=42502436

There has been much debate about exactly how Brunelleschi built the dome without scaffolding (Figs. 3.44, 3.45 and 3.46). He was secretive about his methods and would not allow any record keeping. Likely he had some kind of platform running across the dome, but the details are obscure. Here are some salient observations:

1. The "no-scaffolding" promise was crucial to the bargain, because it was estimated that there was insufficient lumber in Tuscany for the construction of a large enough scaffold.
2. Ghiberti was appointed as co-superintendent of the project. In fact he constantly harassed Brunelleschi and tried to sabotage his work, until Brunelleschi fired him.

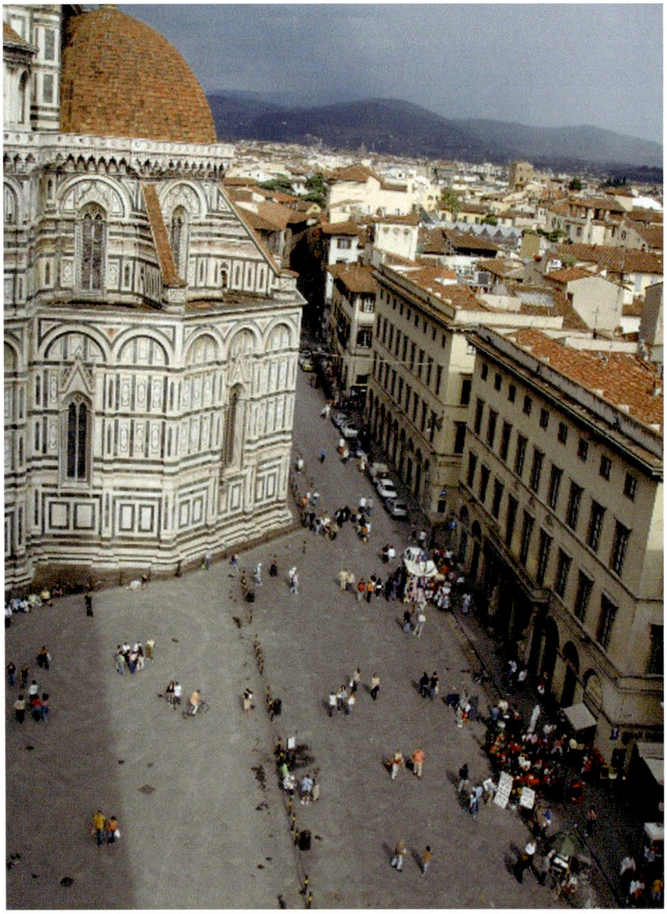

**Fig. 3.46** View of Street from Giotto's Tower—By I, Sailko, CC BY 2.5, https://commons.wikimedia.org/w/index.php?curid=2226099. Florence

3. There are actually two domes, inner and outer (Fig. 3.47).
4. Brunelleschi employed a herring-bone pattern for the brickwork (Fig. 3.48). The bricks are visible to tourists climbing the 463 steps (in the passage between the domes) to the ceiling (Fig. 3.49). (It's worth the climb, for the view of the ceiling, Figs. 2.47 and 2.51. But you will learn that Renaissance men had smaller feet than we do now. I had to ascend sideways in my size-11D clodhoppers.)
5. Of course it was necessary to raise the building materials—and the workers—from the ground to the dome. For this purpose Brunelleschi used an intricate elevator gear system powered by a team of oxen turning an axis. To lower items back to the ground, previous builders using such a system had to unyoke the oxen, turn them around, and yoke them again in order to reverse the action of the elevator. But Brunelleschi's gear accomplished this in minutes, without redirecting the oxen.

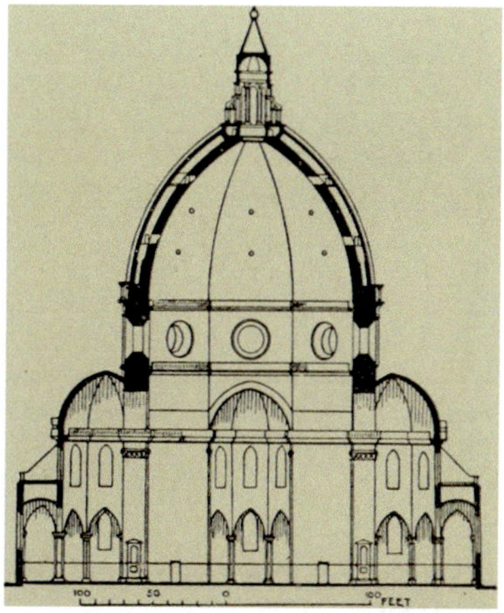

**Fig. 3.47** Nested Domes—Image from page 52 of "The architecture of the renaissance in Italy: a general view for the use of students and others" (1909); Public Domain CC0 1.0 DEED. Internet Archive Book Images https://www.flickr.com/photos/internetarchivebookimages/

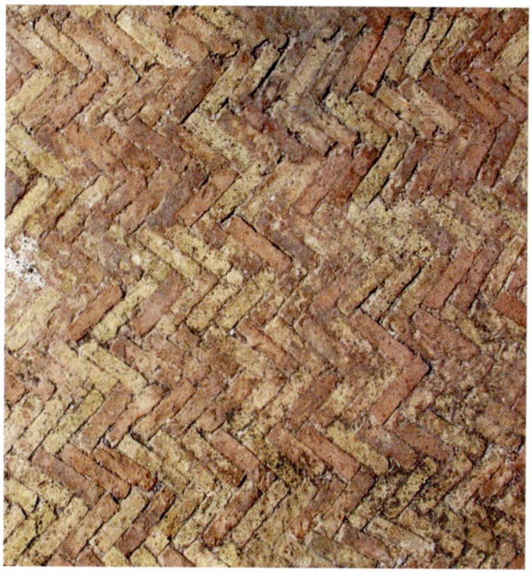

**Fig. 3.48** Herring-Bone Brick Pattern (*opus spicatum*) in Trajan's Market, Rome—Photograph By MM in i.wiki https://it.wikipedia.org/wiki/Utente:MM. Public Domain, https://commons.wikimedia.org/w/index.php?curid=268278

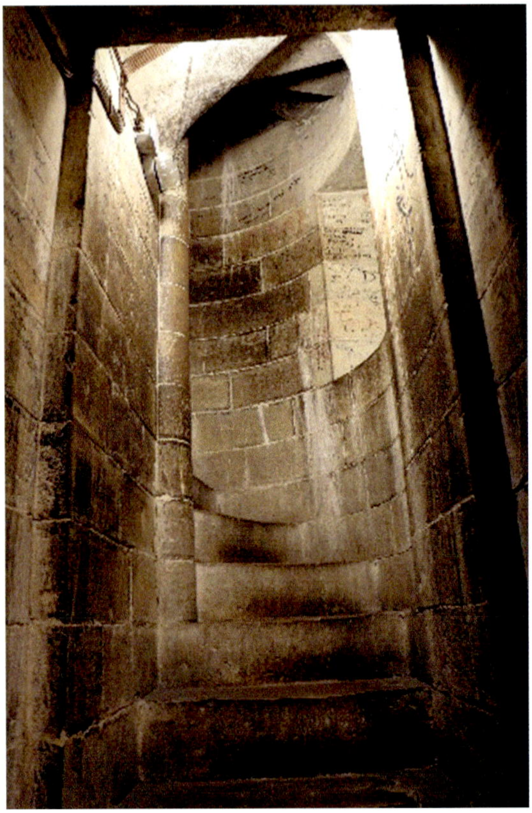

**Fig. 3.49** Florence Duomo Stairs and Inner Passage—By Sailko—Own work, CC BY 3.0, https://commons.wikimedia.org/w/index.php?curid=42377341

6. In 2013 archaeologists uncovered what is suspected to be a builder's model for the dome.
7. In 1989, the architect Massimo Ricci began constructing a 500,000-brick model of the dome in Florence. The experience led to many conjectures as to Brunelleschi's strategies.[6]

All of these details are skillfully, engagingly, and sometimes humorously discussed and illustrated at the following websites. I recommend every one:

https://www.flickr.com/photos/megiess/88398034

https://archinect.com/another/brunelleschi-bim

http://resonanceswavesandfields.blogspot.com/2014/03/brunelleschis-dome-its-structure-and.html

http://www.thehistoryblog.com/archives/22791

[6] Samantha Vaughn: "Massimo Ricci; the man who cracked the Duomo", *The Florentine*, 06 May 2017. https://www.theflorentine.net/2017/05/06/massimo-ricci-man-who-cracked-the-duomo/.

https://www.metalocus.es/en/news/il-duomo-how-filippo-brunelleschi-built-worlds-biggest-dome

https://www.florenceinferno.com/the-brunelleschi-dome/

https://www.archdaily.com/545664/how-did-filippo-brunelleschi-construct-the-dome-of-florence-duomo

https://www.youtube.com/watch?v=_IOPlGPQPuM

https://education.nationalgeographic.org/resource/duomographic/

https://www.princeton.edu/news/2020/05/21/double-helix-masonry-researchers-uncover-secret-italian-renaissance-domes

Two excellent books telling Brunelleschi's story are

*Brunelleschi's Dome: How a Renaissance Genius Reinvented Architecture*, Ross King, Bloomsbury USA; Reprint edition (2013), ISBN-10: 1620401932, ISBN-13: 978-1620401934

and

*Brunelleschi: Studies of His Technology and Inventions*, Frank D. Prager, Gustina Scaglia Courier Corporation, May 24, 2012.

## 3.6 Drawing in Perspective

Let us suppose you want to draw, in perspective, a tiled path receding into the distance (Fig. 3.50).

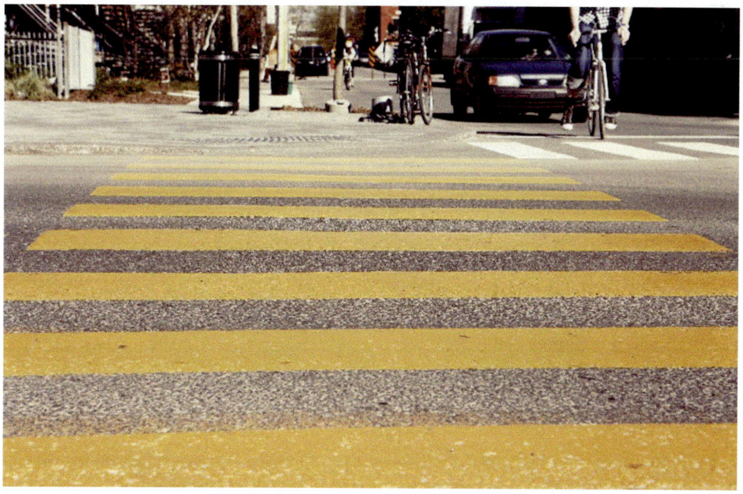

**Fig. 3.50** Pedestrian crossing—By www.Pixel.la Free Stock Photos—crossroad-zebra-crossing-crosswalk, CC0, https://commons.wikimedia.org/w/index.php?curid=51440701

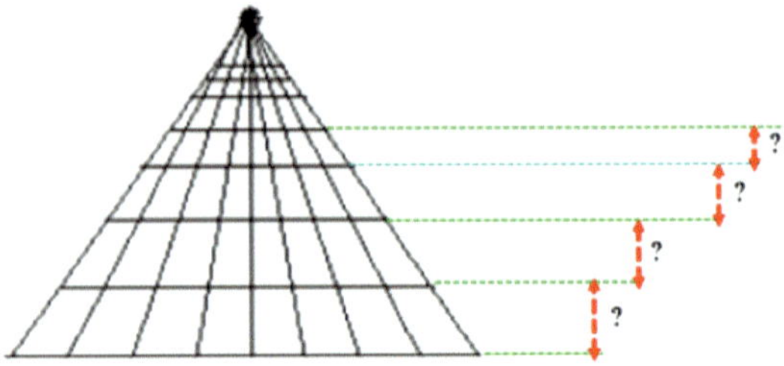


**Fig. 3.51** The transversal problem

How do you determine how far apart, on the canvas, the transversal markings should be (Fig. 3.51)?

The old books describe the construction (Fig. 3.52):

> Draw a trapezoid, subdivide top and bottom equally, draw diagonal, fill in horizontals. … The reconstructed vanishing points of two regions constructed with the zero-vanishing-point method are very unlikely to coincide. (*Zero Vanishing-Point Construction*, Viator (Jean Pélerin), 1505) (see Fig. 3.52)

Got it? No? Would you like to read it again?

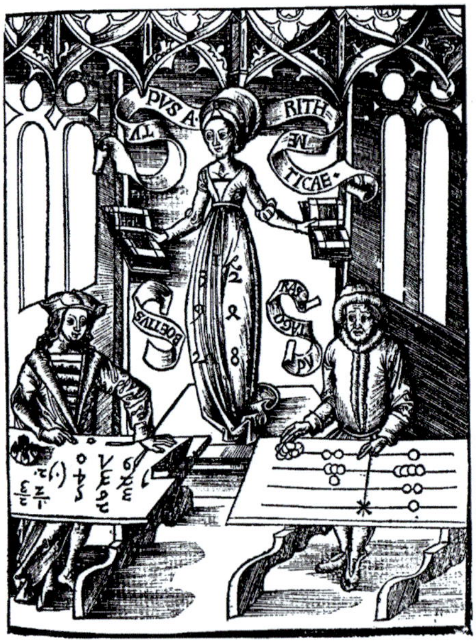


**Fig. 3.52** Drawing from Jean Pélerin's Treatise—By Gregor Reisch (author)—Unknown source, Public Domain, https://commons.wikimedia.org/w/index.php?curid=144963

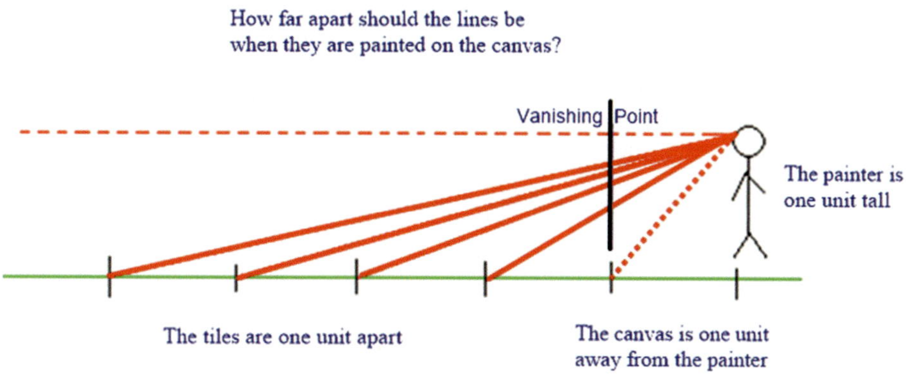


**Fig. 3.53** Working sketch

I didn't think so. We're engineers, and we understand the perspective principle. Let's apply a little analytic geometry to obtain the answer. Figure 3.53 is a working sketch for analyzing Fig. 3.51. The green line is the center line of the path, which we assume runs perpendicular to the painter's canvas. It is marked with horizontal crossings one unit apart; the canvas is situated above one of the markings, and the artist stands one unit back from the canvas. The vanishing point lies where the line of sight, perpendicular to the plane of the canvas, strikes the canvas.

The rightmost (dashed) red line (which does not cross the canvas) has slope 1:1; the second has slope 1:2; the third has slope 1:3, and so on. (This may be clearer if you rotate the page 180°.) *So the lines which intersect the canvas meet it at the distances* 1/2, 1/3, …, $1/n$, … below the vanishing point (Fig. 3.54). As $n$ gets large, the values of this sequence converge to 0—implying, of course, that on the canvas the distances *between* the equispaced transversals are getting smaller and smaller.

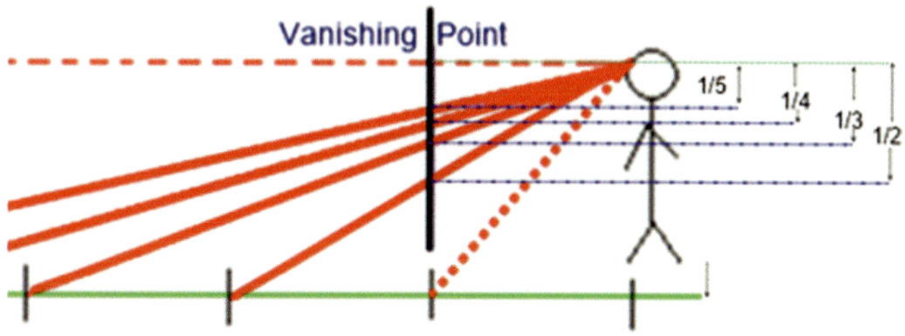


**Fig. 3.54** Crossing locations on the canvas

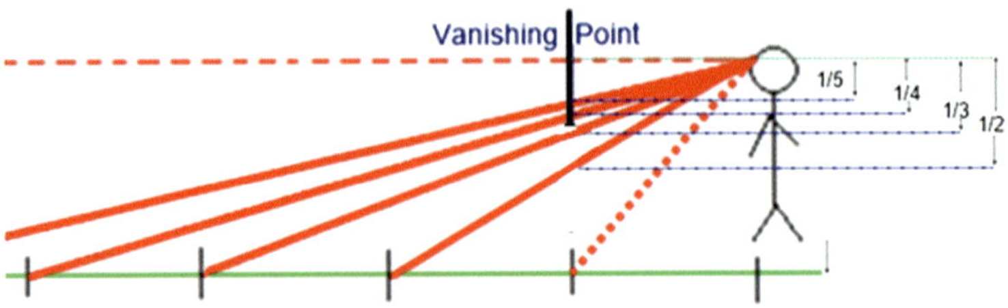


**Fig. 3.55** Crossing locations on a smaller canvas

And of course, the canvas may not be large enough to contain all the intersection points (Fig. 3.55; compare with Fig. 3.54).

Next suppose the path is subdivided into rectangular *tiles*, so that in addition to the horizontal crossings there are equispaced "lane dividers". Will they still be equispaced in the *drawing* (Fig. 3.56)?

**Fig. 3.56** Lane separations

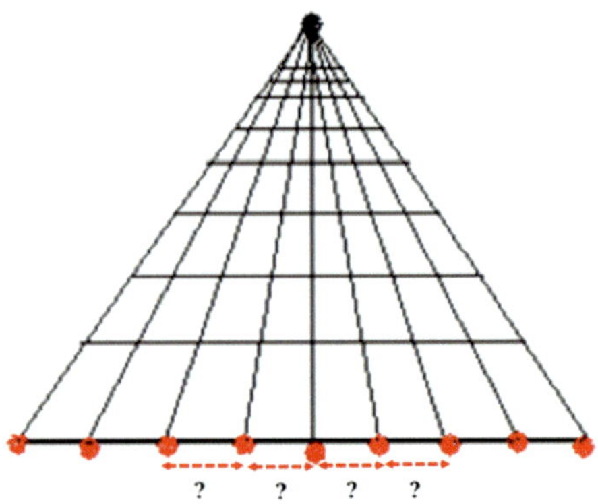

This is best analyzed by looking down from a point above the canvas (Fig. 3.57).

In Fig. 3.57, the green lines are the actual (equispaced) lane markers—separated by *b* units. The red lines are the actual (equispaced) transversals, whose separation *on the canvas* we just worked out. The blue segment is the canvas and the grey lines are the lines of sight from the painter's eye to the points where the lane markers intersect the transversal located at $x = a$. Note that the grey lines have the equations $y = 0, y = \pm\frac{b}{a}x, y = \pm 2\frac{b}{a}x, \ldots$. They intersect the canvas (where $x = c$) at $y = 0, y = \pm\frac{b}{a}c, y = \pm 2\frac{b}{a}c, \ldots$. *The lane markers remain equispaced in the painting*!

So: how can an artist draw tiled paths in perspective? Study Figs. 3.58, 3.59, 3.60, 3.61, 3.62, 3.63 and 3.64.

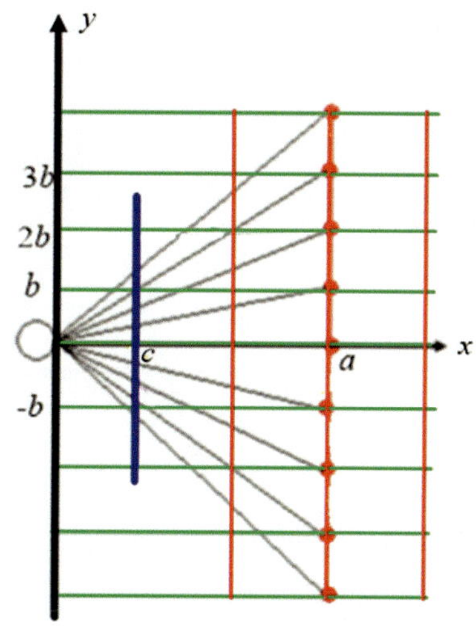


**Fig. 3.57** Sketch for analyzing lane separation

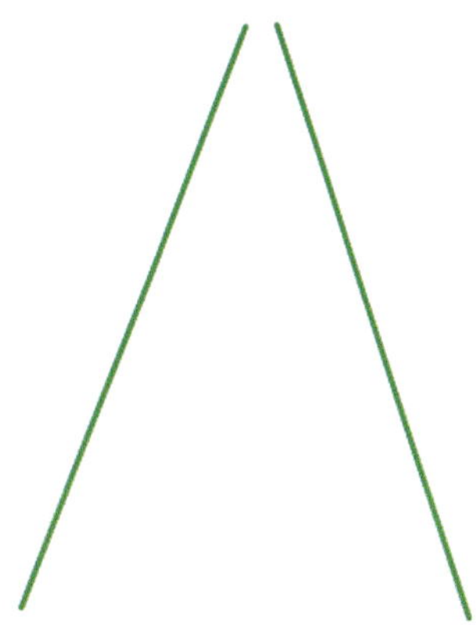

**Fig. 3.58** Draw the edges of the path

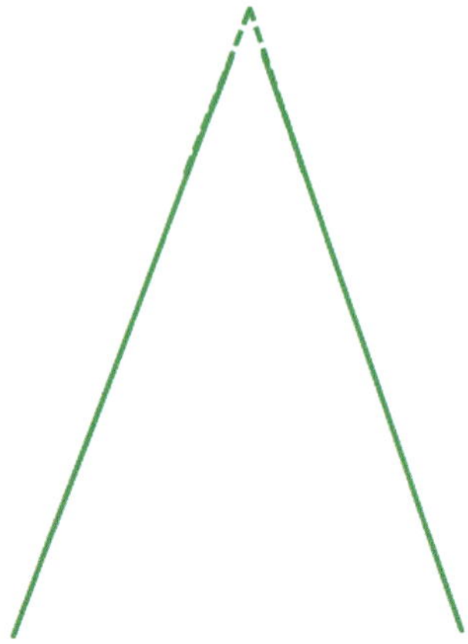

**Fig. 3.59** Determine the vanishing point

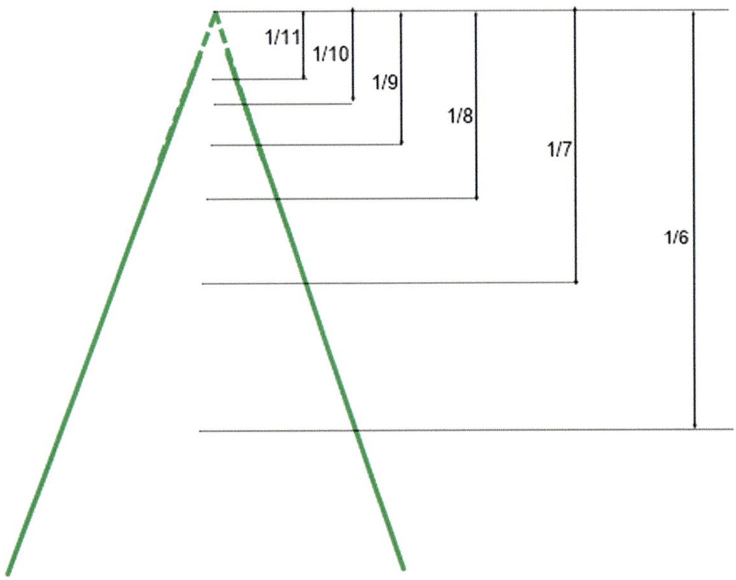


**Fig. 3.60** Use the $1/n$ rule to specify the transversal locations[7]

[7] $1/N$ times … what? *Times the vertical height of the vanishing point*. Figure 3.53 reminds us that we have made many assumptions to simplify our presentation: the path runs directly away from the canvas, the height of the vanishing point equals the height of the (eye of the) artist equals the transversal spacing; the artist and the canvas are standing on transversals; etc.

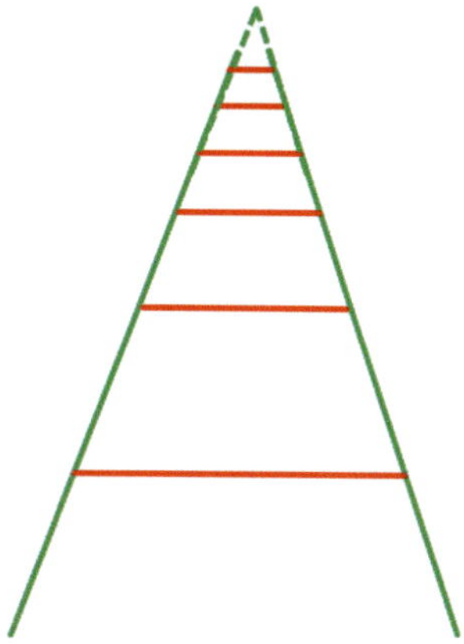

**Fig. 3.61** Draw the transversals

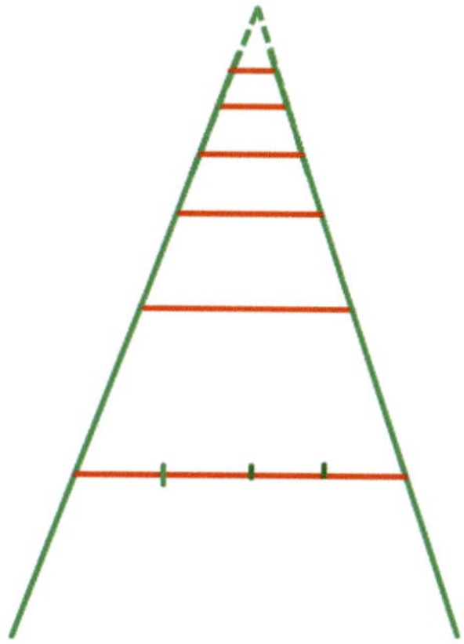

**Fig. 3.62** Mark equal spacings for the lanes

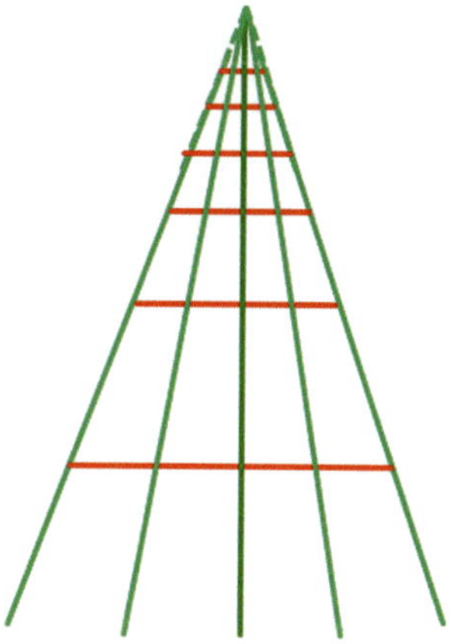

**Fig. 3.63** Draw the lane markings

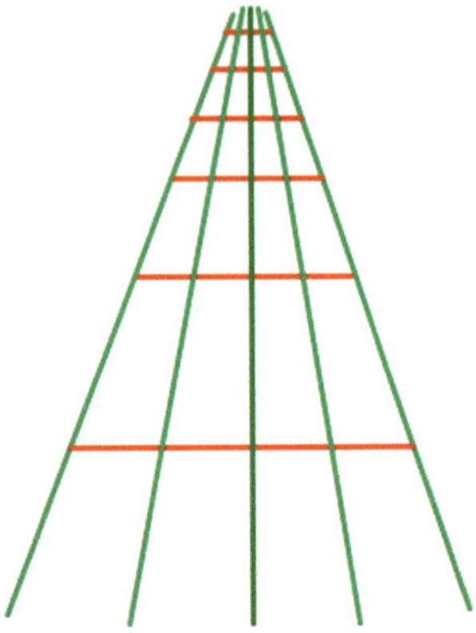

**Fig. 3.64** Erase the construction lines

## 3.7 Impossible Objects

The chief objective of the theory of perspective is the depiction of three-dimensional objects on a two-dimensional medium (the canvas). We have explored this through the postulate that perspective is achieved when light coming from each point on the object enters the eye from the same direction as light from the corresponding point on the canvas image; recall Durer (Fig. 3.7). This is called "Linear Perspective" by artists. In Chap. 5 we'll look at another approach to perspective.

But for now we'll going to see how some artists have had a little fun with linear perspective and created drawings that, when viewed on small scales, appear to accurately represent parts of real (3D) objects; but when visualized on a wider—well, perspective—they are connected in impossible combinations. You'll probably recognize the devil's fork, the Penrose triangle, the Escher stairwell, the Necker cube, and the impossible cube (Figs. 3.65, 3.66, 3.67, 3.68 and 3.69).

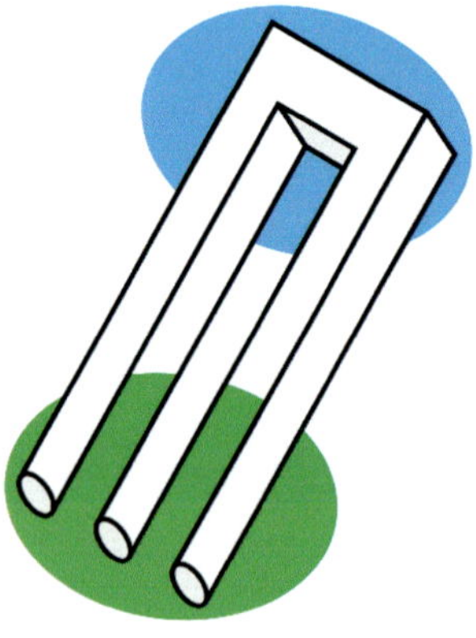

**Fig. 3.65** Devil's Fork or Impossible Trident—By AnonMoos—This vector image includes elements that have been taken or adapted from this file, Public Domain, https://commons.wikimedia.org/w/index.php?curid=12680203. https://en.wikipedia.org/wiki/Impossible_trident#/media/File:Poiuyt--opaque.svg

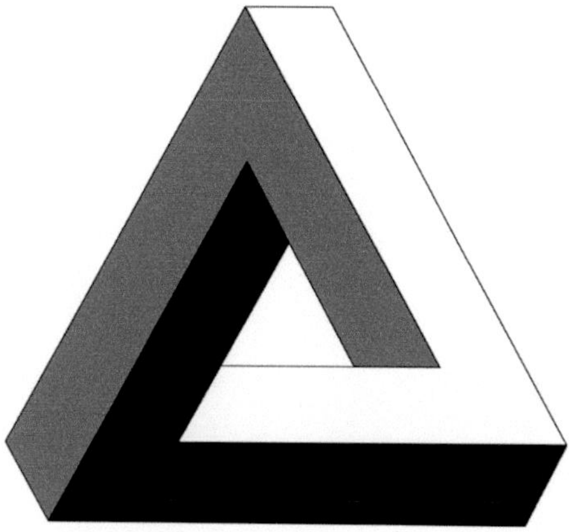

**Fig. 3.66** Penrose Triangle—By Tobias R.—Metoc—Own work, Public Domain, https://commons.wikimedia.org/w/index.php?curid=2520370. https://en.wikipedia.org/wiki/Penrose_triangle#/media/File:Penrose-dreieck.svg

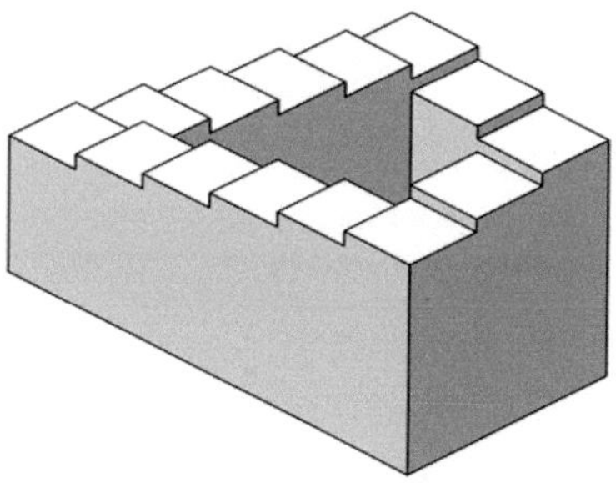

**Fig. 3.67** Escher Stairwell—By Sakurambo—Own work, Public Domain, https://commons.wikimedia.org/w/index.php?curid=794844. https://commons.wikimedia.org/wiki/File:Impossible_staircase.svg?uselang=en#/media/File:Impossible_staircase.svg

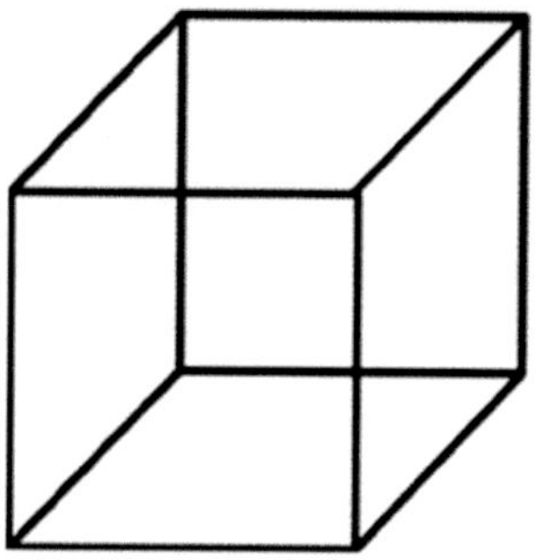

**Fig. 3.68** Necker Cube By No machine-readable author provided. Albert kok assumed (based on copyright claims). Own work assumed (based on copyright claims). Public Domain https://commons.wikimedia.org/w/index.php?curid=1645066. Cropped from the original

The devil's fork was discovered by D. H. Schuster in 1964. The Penrose triangle was conceived by Oscar Reutersvärd (1934) and later popularized by Lionel and Roger

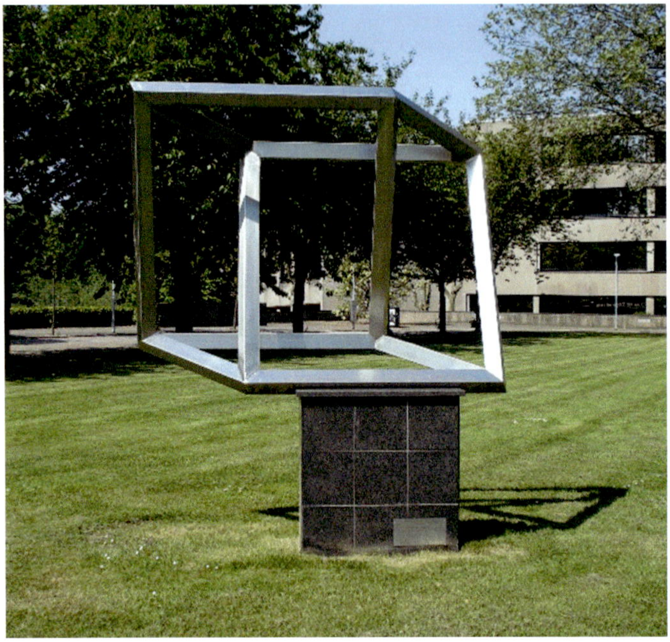

**Fig. 3.69** *The Impossible Cuboid* (**Verhoeff**, **Baas Becking**, and **Bakker**)—By Wikifrits—Own work, Public Domain, https://commons.wikimedia.org/w/index.php?curid=4173815. 1988; Universiteit Burgemeester Oudlaan, Rotterdam

Penrose (1958). The Escher stairwell was also conceived by Reutersvärd (1937) and popularized by the Penroses, but came into the public eye through its inclusions in the drawings *Klimmen en dalen* (*Ascending and Descending*)[8] (1960) and *Waterval* (*Waterfall*)[9] (1961) by Maurits Cornelis Escher. The Necker cube was conceived in 1832 by L. N. Necker and is included in Escher's *Belvedere*[10] (1958) together with his impossible cube. Naturally these gizmos go by many other names.

But the granddaddy of impossible perspective is the old 1753 chestnut by Hogarth. The original version can be seen at https://commons.wikimedia.org/w/index.php?curid=644038. Figure 3.70 displays an engraving executed by T. Cook in 1754. Scrutinize it in detail and be prepared to laugh.

[8] https://en.wikipedia.org/wiki/ascending_and_descending.

[9] https://en.wikipedia.org/wiki/Waterfall_(M._C._Escher).

[10] https://en.wikipedia.org/wiki/Belvedere_(M._C._Escher).

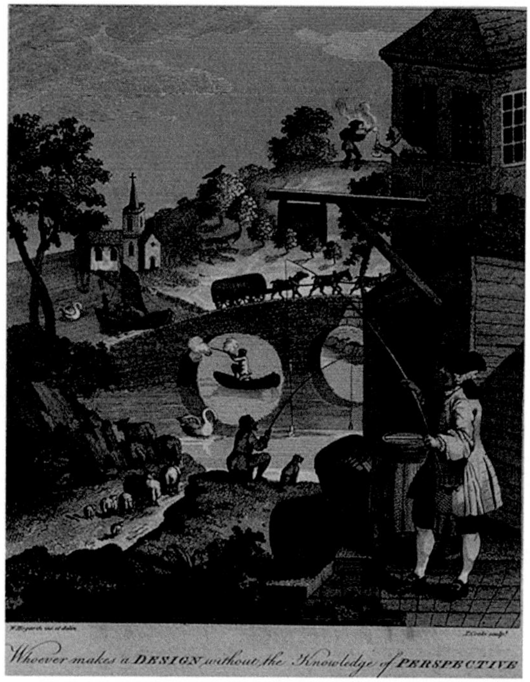


**Fig. 3.70** *Satire on false perspective* (Cook after Hogarth)—By https://wellcomeimages.org/indexplus/obf_images/49/be/4e7fb1e85c7a6b4b86e801a543c4.jpgGallery: https://wellcomeimages.org/indexplus/image/V0049303.html Wellcome Collection gallery (2018-04-02): https://wellcomecollection.org/works/sep5e7sr CC-BY-4.0, CC BY 4.0, https://commons.wikimedia.org/w/index.php?curid=36668131. Cropped from the original. 1754

Escher (1898–1972) was a Dutch graphic artist whose works have delighted and confounded mathematicians for years. He became a mathematical celebrity when attendees at a meeting of the International Congress of Mathematicians in 1954 in Amsterdam discovered his works at an exhibition in the Stedelijk Museum. Explore his "takes" on perspective, tilings of the plane and their symmetries, and visualizations of infinity at the following sites:

https://mcescher.com/

https://en.wikipedia.org/wiki/M._C._Escher

https://mcescher.com/gallery

The website http://im-possible.info/english/index.html is full of impossible figures devised by Escher and others, and contains an excellent article by Bruno Ernst, "Adventures with Impossible Figures": https://im-possible.info/english/articles/awif/index.html. The authors cited in the following reading list represent some of the mathematicians, physicists, and crystallographers who influenced and/or collaborated with Escher:

Coxeter, H. S. M. (1979), "The non-Euclidean symmetry of Escher's picture 'Circle Limit III'", *Leonardo* ***12****: 19–25.*

Haag, Friedrich, "Die regelmäßigen Planteilungen und Punktsysteme," *Zeitschrift für Kristallographie*, Volume 58, 1923, pp. 478–489.

MacGillavry Carolina H., "Symmetry Aspects of M. C. Escher's Periodic Drawings," *IUCr*, Utrecht, 1965.

Penrose, Lionel S.; Penrose, Roger (1958). "Impossible objects: A special type of visual illusion," *British Journal of Psychology* ***49***: *31–33*.

Pólya, George (1924) "Über die Analogie der Kristallsymmetrie in der Ebene," *Zeitschrift für Kristallographie*, vol. 60, pages 278–282.

Robinson Sara, "M. C. Escher: More Mathematics Than Meets the Eye," *SIAM News*, Volume 35, Number 8, October 2002.

Schattschneider, Doris, *M. C. Escher: Visions of Symmetry*, W. H. Freeman, 1990, new edition Harry N. Abrams, 2004.

Schattschneider, Doris, "The Mathematical Side of M. C. Escher," *Notices of the AMS* Volume 57, Number 6, 706–718 (2010).

Yang, Chen Ning, "Symmetry Principles in Physics," *Phys. Teach.* 5, 311 (1967).

Escher's regular-polygon tilings have lots of examples based on triangles, squares, and hexagons (Figs. 3.71, 3.72 and 3.73).

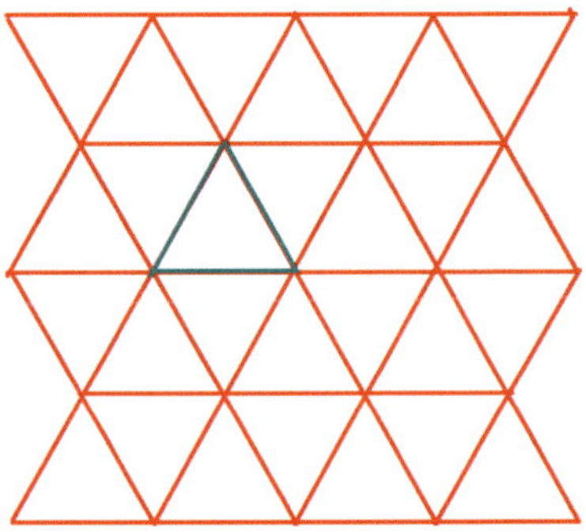

**Fig. 3.71** Triangular tiling

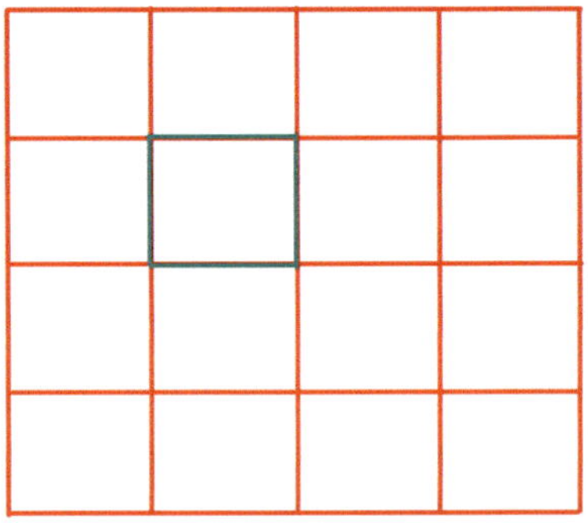

**Fig. 3.72** Square tiling

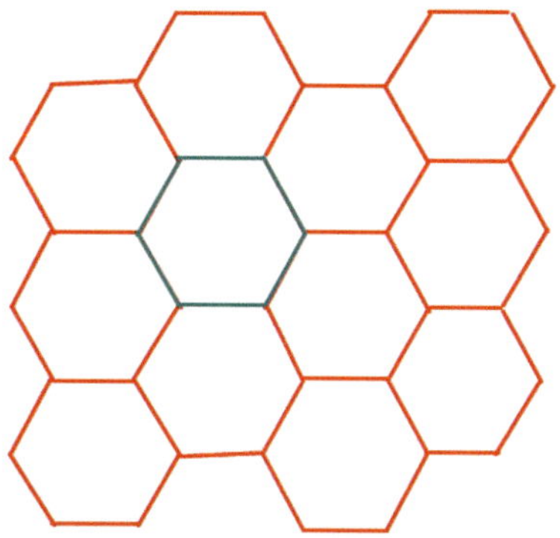

**Fig. 3.73** Hexagonal tiling

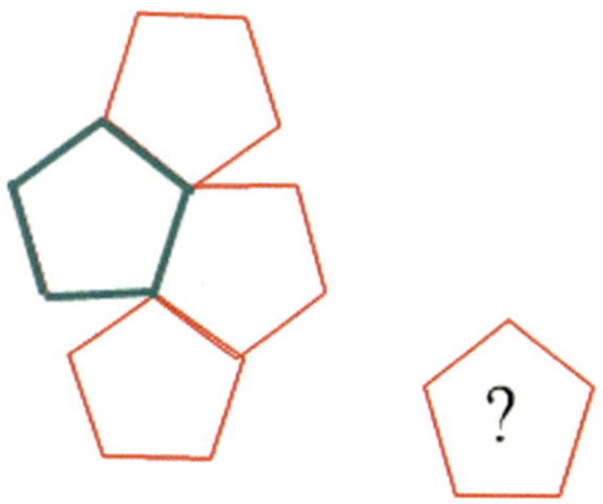


**Fig. 3.74** Pentagonal Ti… never mind

Why not pentagonal, heptagonal, or octagonal (Fig. 3.74)?

You might have fun with this exercise.

EXERCISE. Show that there are only three tilings of the plane with regular polygons, namely, with triangles, squares and hexagons.

(Hint: Note from Fig. 3.75 that if you walk around the perimeter of a regular $n$-gon you make $n$ left turns through (180–θ) degrees, for a net turn of 360°. Conclude that the formula for the interior angle of a regular $n$-gon is $\theta = \frac{180n-360}{n}$. If the tiling fills the plane then θ must divide 360°: $\theta = \frac{180n-360}{n} = \frac{360}{m}$ or $\frac{2n}{n-2} = m$ for some integer $m$. Plot $\frac{2n}{n-2}$ for $n = 3, 4, 5, 6, 7, 8, \ldots$ and you'll see the answer.)

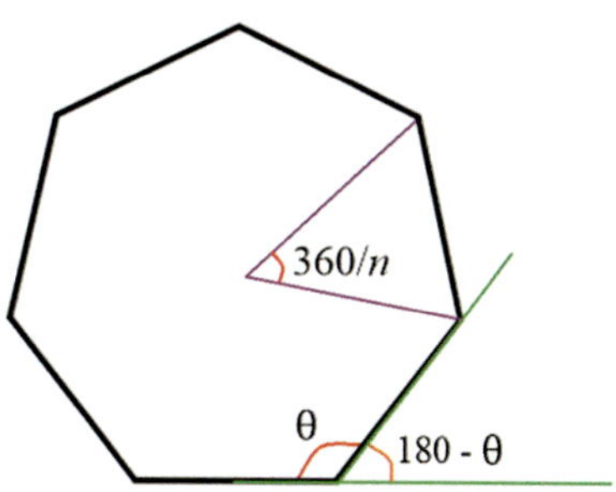


**Fig. 3.75** Regular $n$-gon

# 4 Color Theory: Physics and Physiology

## 4.1 What is Light?

Many of our idioms are premised on an ancient, incorrect understanding of how we see.

"Cast a glance at this."

"Feast your eyes on that."

"Can't take my eyes off of you."

"I could feel him undressing me with his eyes."

These all suggest that our eyes probe for information by emitting some sort of wave. Several comic book superheroes have reinforced this misconception (Fig. 4.1).[1]

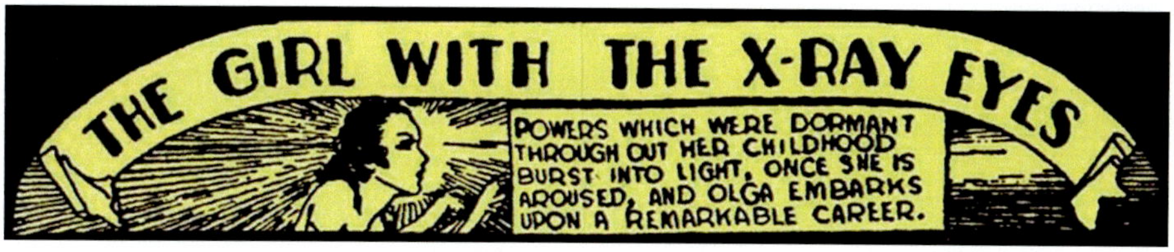


**Fig. 4.1** Comic book eyes emitting radiation - By (Original text: Logo, artist unknown, from a magazine published between 1923 and 1963 (1937 or 1938) with copyright not renewed: Public domain source: scan of a 1937 or 1938 issue of spicy mystery stories at http://www.oocities.org/pippl2/olga01.jpg), Public domain, https://commons.wikimedia.org/w/index.php?curid=54703872

[1] Creative Commons License Links for all images are listed in the Preface.



A. D. Snider, *From STEM to STEAM*, Synthesis Lectures on Engineering, Science, and Technology, https://doi.org/10.1007/978-3-031-57316-3_4

So let's be sure we have it right. *We "see" an object when a light source illuminates the object and it reflects some of this light into our eyes.* (Proof? If we turn off the light source, we don't see the object.) Of course we can also look directly into the source. (*Seldom advised.*)

What are the features of this light radiation? The simplest property is its speed—but it's hard to measure, because it's so fast.

## 4.2 The Speed of Light

The French physicist Armand Hippolyte Louis Fizeau (1819–1896) contrived a very clever experiment to measure the speed of light $c$. The basic idea behind his design involves a light source situated behind a screen with a small aperture; the light beam coming through the aperture would propagate to a remote mirror, get reflected, and return through the aperture. *However*, suppose he placed a *very fast* shutter over the aperture, which opened to send the beam toward the mirror but subsequently closed so rapidly that it—just barely—blocked the returning beam. Then there would be no reflected light; and he could determine the light's velocity by dividing the screen-to-mirror distance $d$ (times two) by the length of time that the shutter was open.

Fizeau needed to design a very fast shutter, delay the time-of-travel of the light beam, and separate the reflected beam at the screen from the source's emitting light (so he can tell when the reflection has been blocked). Here's a schematic of his apparatus (Fig. 4.2).

Study the diagram (and ignore the lenses, which only focus and collimate the light). A continuous beam of light is emitted by the source. Part of the beam is transmitted *through* the (semitransparent) mirror; that part does not concern us. The part that is *not* transmitted is reflected and directed down the long path to the left.

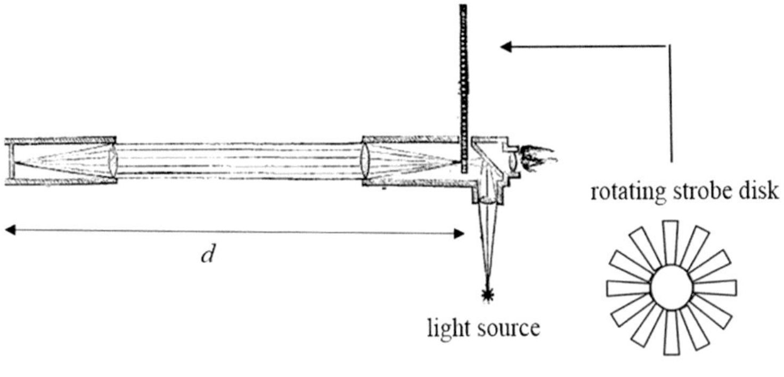


**Fig. 4.2** Fizeau's Apparatus—*The manufacturer and builder*, volume 10 issue 11 (november 1878), page 253. Public Domain. Courtesy of HathiTrust: https://babel.hathitrust.org/cgi/pt?id=coo.31924080795333&seq=259. Cropped and embellished from original

If the disk is stationary and its blades do not block the beam, then the left-reflected beam travels the distance $d$ and is reflected again, by the remote mirror on the far left. It returns to the original mirror and once again is partially reflected back into the source; we ignore that part. The other part is transmitted into the eye. The eye sees a continuous stream of light.

If a blade of the disk *does* block the beam, of course no light makes it to the eye; we see darkness. From this blocking position, if the disk is *slowly* rotated, at some instant the blade ceases to block the beam and, $2d/c$ seconds later, the eye sees the reflected light. As the disk continues to rotate, we see a strobe light effect.

But if we speed up the disk rotation so that after $2d/c$ seconds, the *next* disk blade comes into position and blocks the returning beam, then the eye never sees the light that made the round trip; it sees darkness. To sum up: if the disk spins slowly at first but then is accelerated, the eye sees light, dark, light, dark, ..., alternating until the disk reaches the critical speed and the eye sees no light.

How long is the aperture open? Let $n$ be the number of blades in the disk, and assume that the blades and slots are evenly spaced. Then the angle occupied by one slot is $360/(2n)$ degrees. If the disk turns at $r$ revolutions ($360r$ degrees) per second, the time required for a blade to occupy the position of an adjacent slot is $\frac{360/2n}{360r} = \frac{1}{2nr}$ seconds. Fizeau's disk had $n = 720$ teeth (!), enabling him to achieve very high shutter speeds.

His separation distance $d$ was 5.25 miles (!!). The semitransparent mirror provided the isolation of the reflected beam from the source light. Fizeau observed that the reflection was completely blocked when the disk speed $r$ reached 12.6 rev/s, leading him to conclude that

$$c = \frac{2d}{1/2nr} = 4ndr = 4 \times 720 \times 5.25 \times 12.6 = 190{,}512 \text{ miles per second.}$$

EXERCISE. If Fizeau doubled the disk rotation speed to 26.2 rev/s, what would he see?

Modern measurements set the speed of light at 186,282 miles per second or 299,792,458 m per second.

## 4.3 Wavelength, Frequency, and Color

A considerable amount was known about light even before the elegant electromagnetic equations of James Clerk Maxwell (1831–1879) were formulated. Isaac Newton (1642–1726) and his colleagues deduced from the familiar prism refraction phenomenon depicted in Figs. 4.3 and 4.4 that white light was made up of a superposition of colored light rays:

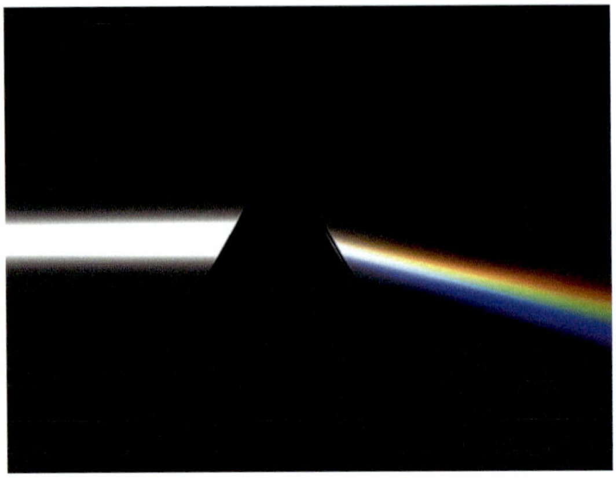

**Fig. 4.3** Photograph of refraction through a prism—by fdecomite CC BY 2.0 DEED. https://www.flickr.com/photos/fdecomite/8530626099

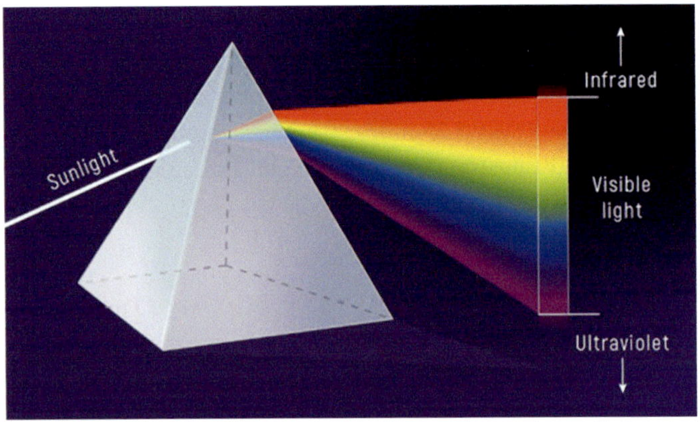


**Fig. 4.4** Schematic of prism refraction—NASA, European Space Agency, Leah Hustak (Space Telescope Science Institute). Public domain. https://webbtelescope.org/contents/media/images/01F8GF658T060QGSSZYR1GD6GA

And interference experiments with the colored rays revealed that they were sine waves (Figs. 4.5 and 4.6).

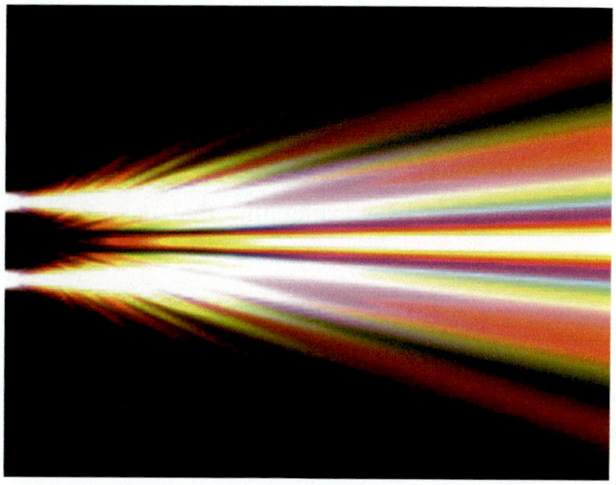

**Fig. 4.5** Double slit interference pattern—By Timm Weitkamp (talk)—Own work, CC BY 3.0 de, https://commons.wikimedia.org/w/index.php?curid=14919517

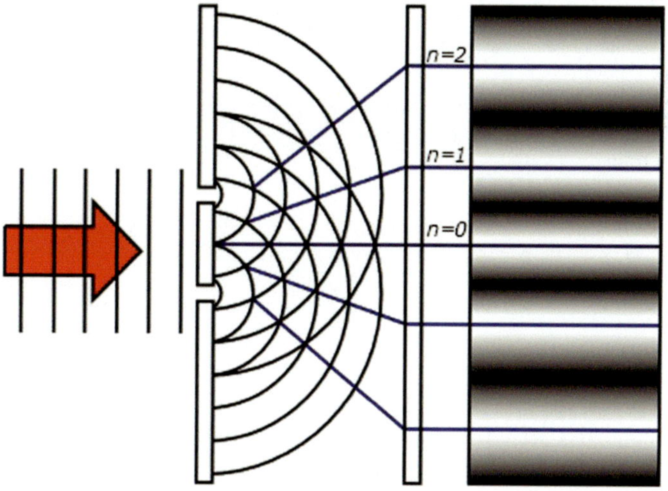


**Fig. 4.6** Destructive/constructive interference—By inductiveload—Own work (Own drawing), Public domain, https://commons.wikimedia.org/w/index.php?curid=1663904. Cropped and embellished from original

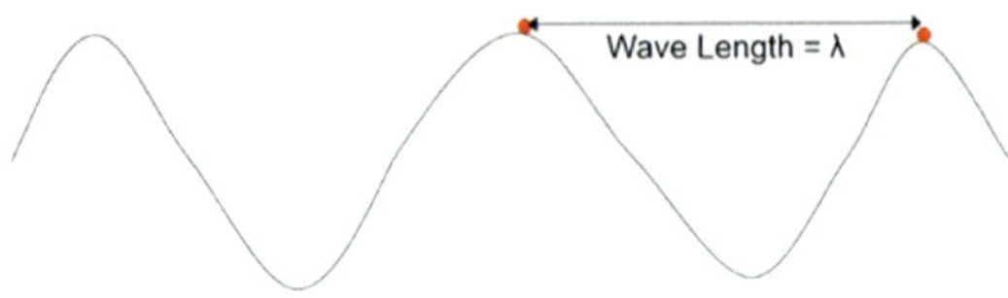


**Fig. 4.7** Wavelength

A sine wave is characterized by its wavelength, illustrated in Fig. 4.7, depicting a "freeze photograph" of the wave at one instant:

In one second the sine wave will have traveled the distance $c$ (Fig. 4.8),

So its *frequency* $f$, the number of wavelengths that pass a given fixed point in one second, is

$$f = c/\lambda \text{wavelengths per second or Hertz(Hz)}.$$

The reciprocal of frequency, $T = 1/f$, is the *period* of the oscillations at the fixed point, i.e. the time lapse for one complete oscillation at the point (Fig. 4.9):

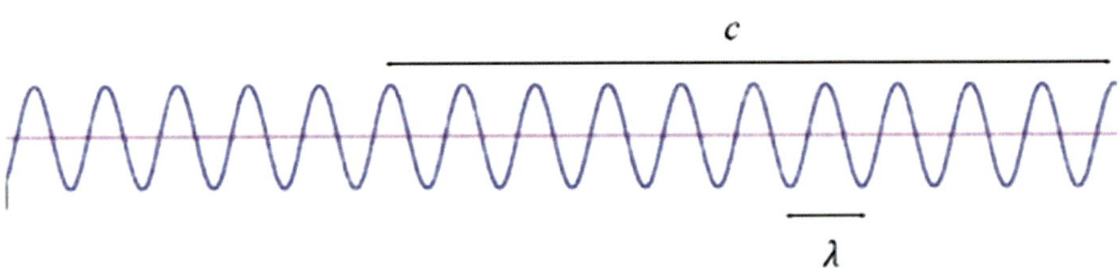


**Fig. 4.8** Wave train

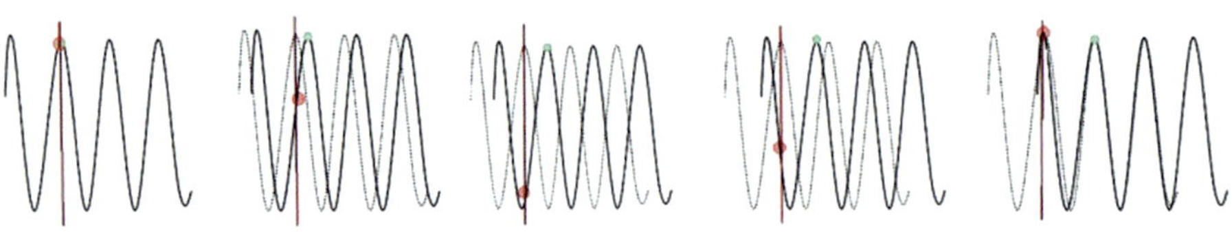

**Fig. 4.9** One period

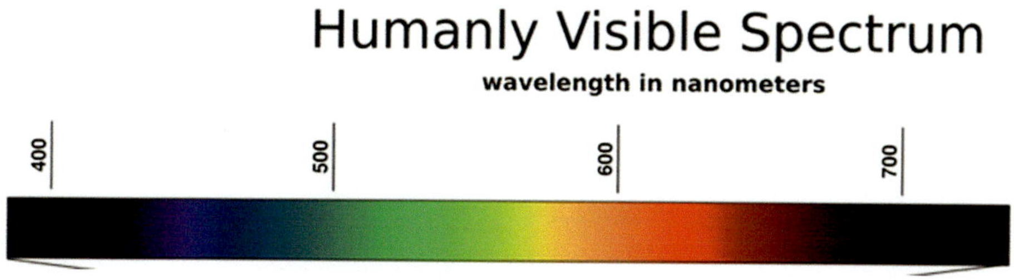


**Fig. 4.10** The visible spectrum—By Johannes Ahlmann. CC BY 2.0 DEED. https://www.flickr.com/photos/entirelysubjective/6146868258. Cropped from original

Visible light is a very high-frequency, short-wavelength phenomenon. A typical wavelength of red light is 680 nm = $6.8 \times 10^{-7}$ m, and its frequency is $4.41 \times 10^{14}$ Hz. For green light a typical wavelength is 545 nm and the frequency is $5.50 \times 10^{14}$ Hz. See Fig. 4.10.

## 4.4 Refraction

The prism experiment illustrates the phenomena of refraction. The speed of light is slower in the prism than in air, and "refraction" refers to the bending of light rays at boundaries where the propagation speed changes. Refraction is explained quite rigorously by Maxwell's equations, but we can visualize it as follows. Imagine you are driving a car and you are about to cross a line where the road surface changes from asphalt to gravel, reducing your speed. If the interface is skewed with respect to the road as in Figs. 4.11, 4.12 and 4.13, the car's right wheel will be slowed down before the left, resulting in a turn in its trajectory.

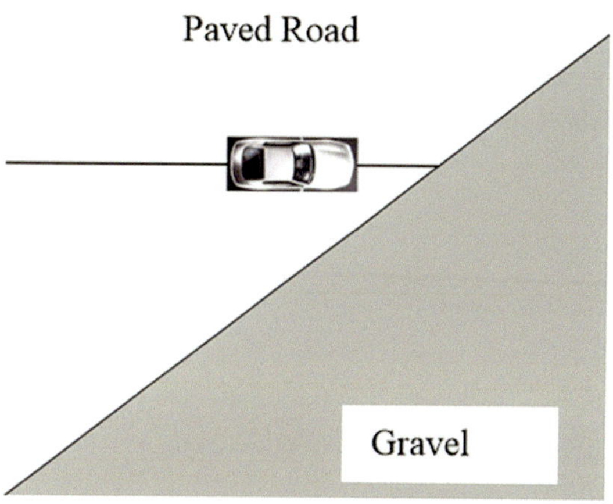


**Fig. 4.11** Crossing to a slower medium: the approach (**Joshua Nicodemus**) public domain

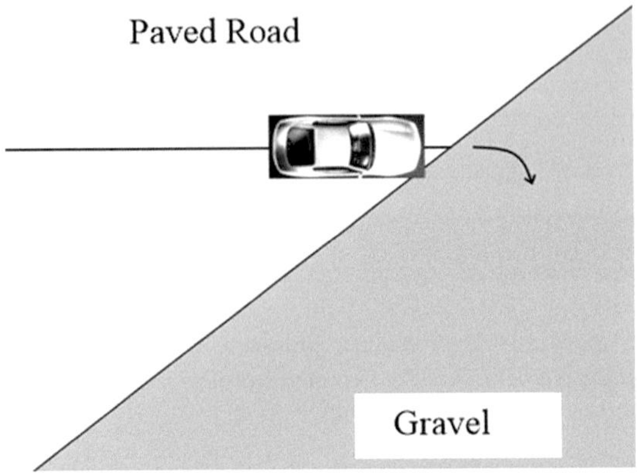


**Fig. 4.12** Crossing to a slower medium: crossing the interface (**Joshua Nicodemus**) public domain

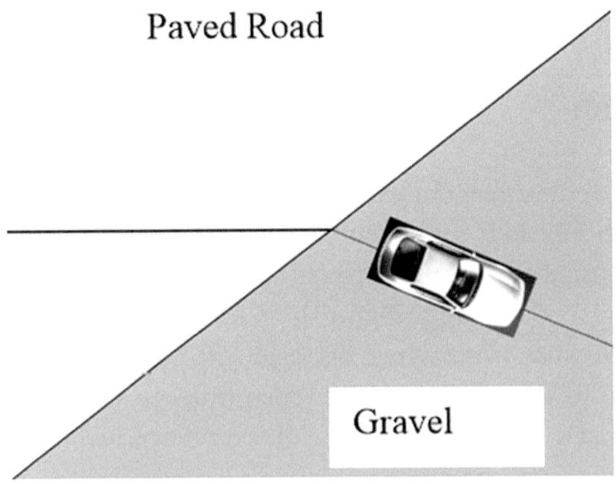


**Fig. 4.13** Crossing to a slower medium: continuing (**Joshua Nicodemus**) public domain

When the car crosses back onto the asphalt, its trajectory is bent again because the left wheel speeds up first (Figs. 4.14, 4.15 and 4.16).

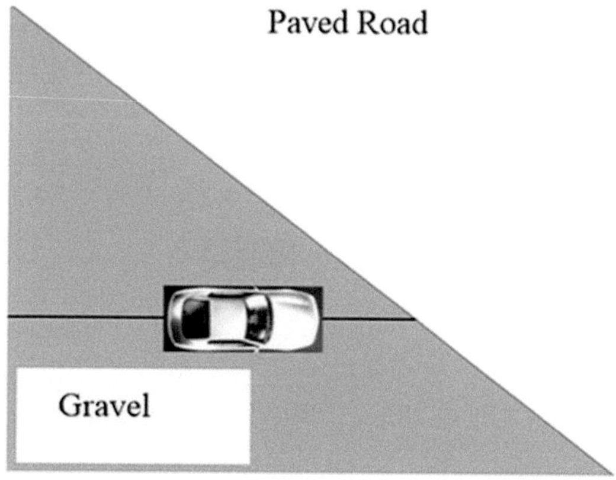


**Fig. 4.14** Crossing to a faster medium: the approach (**Joshua Nicodemus**) public domain

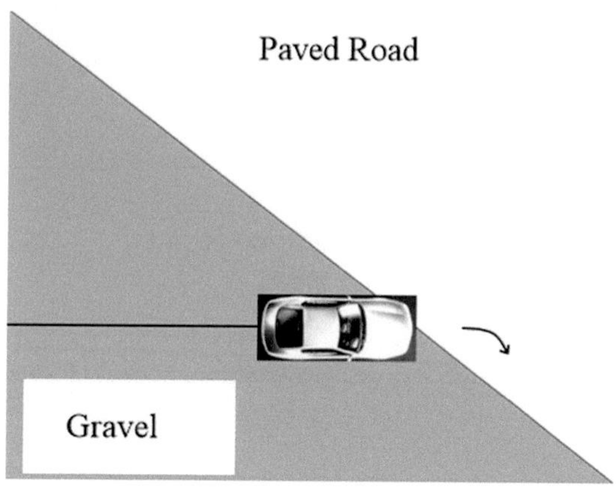


**Fig. 4.15** Crossing to a faster medium: crossing the interface (**Joshua Nicodemus**) public domain

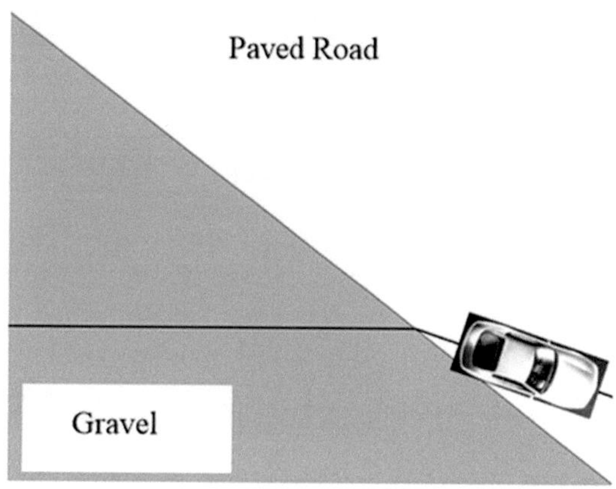


**Fig. 4.16** Crossing to a faster medium: Continuing (**Joshua Nicodemus**) public domain

Analyze the bending of the car's trajectory for the geometries in the figures below, and confirm that *in all cases the car's trajectory bends* **towards** *the normal to the interface when passing from a high speed medium to a low speed medium, and* **away** *from the normal otherwise* (Figs. 4.17 and 4.18).

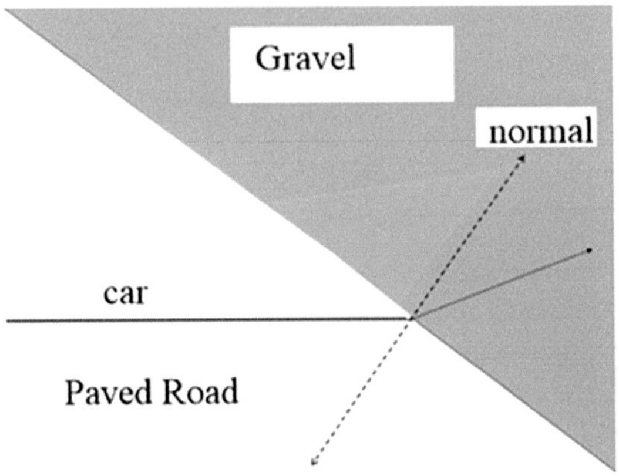


**Fig. 4.17** Trajectory bends towards the normal (**Joshua Nicodemus**) public domain

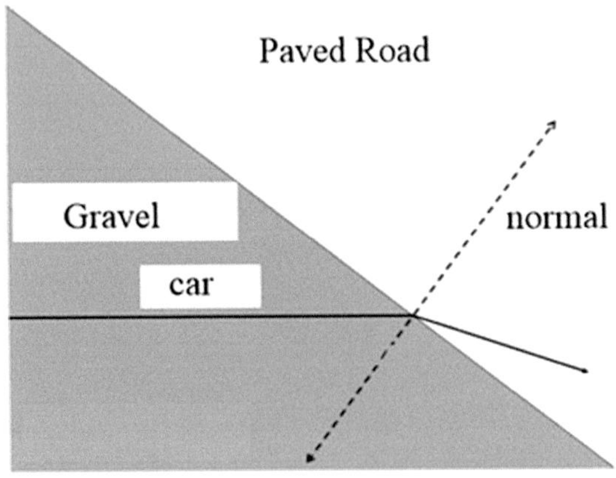


**Fig. 4.18** Trajectory bends away from the normal (**Joshua Nicodemus**) public domain

Light beams refract according to the same rule, so if they enter *and exit* a prism the situation resembles Fig. 4.19.

Note that if the prism is shaped like a diamond, diverging beams are bent into converging beams. This suggests the "refocusing" action of a *lens* (Fig. 4.20).

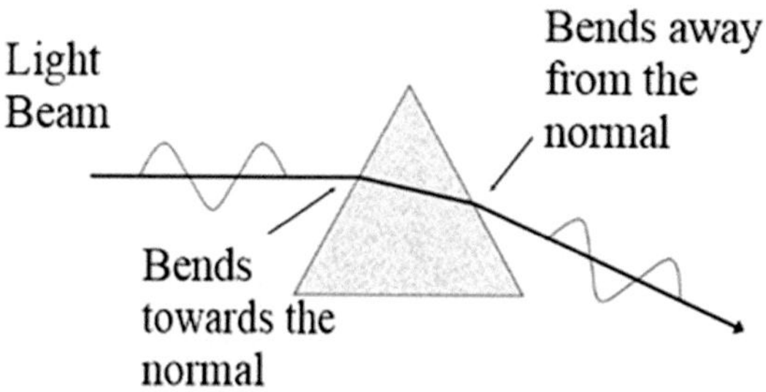


**Fig. 4.19** Diffraction through a prism

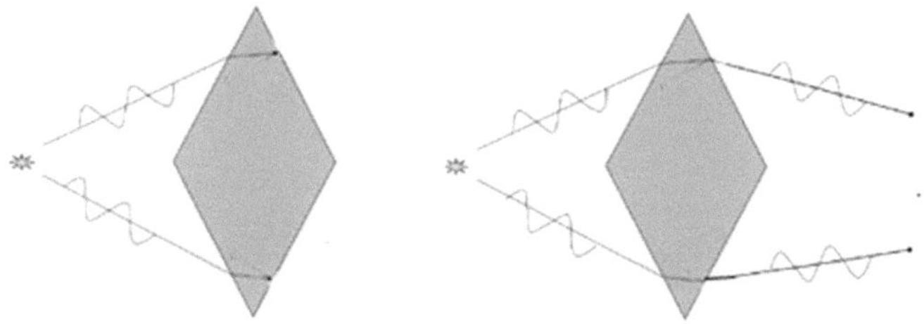

**Fig. 4.20** Refocusing: Incoming beam bends toward the normal and outgoing beam bends away from the normal

The bending of light at an interface has been quantified experimentally and theoretically via *Snell's law*: in Fig. 4.21 if $v_i$ is the wave velocity and $n_i \equiv c/v_i$ is the refractive index in material $i$,

$$\textbf{Snell's Law} \quad \frac{n_1}{n_2} = \frac{\sin\theta_2}{\sin\theta_1}$$

EXERCISE. Using the angles $\theta_1$ and $\theta_2$, express the wave velocity vector in each medium in terms of its component *normal* to the interface and its component parallel to the interface. Then show that Snell's law implies that the velocity components parallel to the surface are equal.

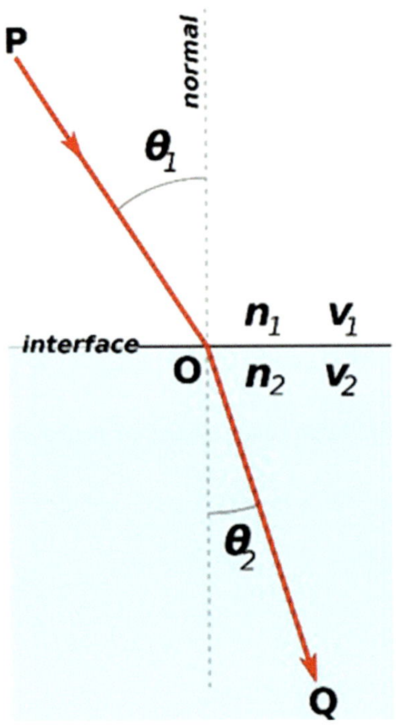


**Fig. 4.21** Snell's law

## 4.5 Dispersion

Observe from the prism experiment (Figs. 4.3 and 4.4) that refraction separates the various colors that make up a light beam, *by bending them through different angles*. The same phenomenon occurs inside raindrops, producing the rainbow. In the auto/highway analogy, a softer gravel surface will slow the wheels down more, resulting in a sharper turn in the trajectory. So we can explain the separation of colors by postulating that the speed of light in the prism or raindrop changes according to the light's color. "Dispersion" is the name of this effect.

It was experimentally determined by scholars around the thirteenth century that a key feature in the formation of a rainbow is the fact that sunlight enters atmospheric water droplets and becomes reflected off the back surface. In Fig. 4.22 the light beam undergoes refraction upon entering and exiting the droplet, and thus its colors are separated and spread—twice, as with the prism. The details of the positioning and shape of the rainbow are fairly complicated matters that do not concern us here.

EXERCISE. Take a photo of a rainbow and verify the order of the appearance of the colors is as depicted in the exit beam of Fig. 4.22.

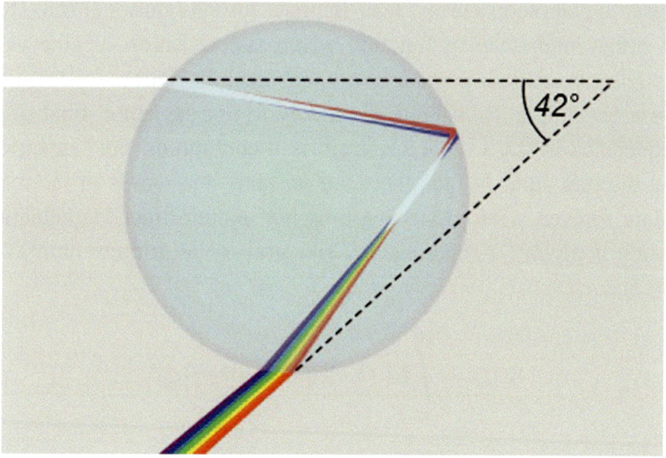


**Fig. 4.22** Refraction/reflection/refraction of a sunbeam—By KES47—Own work, public domain, https://commons.wikimedia.org/w/index.php?curid=10636870

Since light speed varies with frequency, we need to revisit the frequency rule $f = c/\lambda$. In vacuum and air, the speed of light $c$ is constant, independent of frequency. But when we hit a dispersive medium, $c$ changes with frequency. So either the frequency or the wavelength, or both, must change—which?

Not the frequency! The number of times the wave hits its peak in one second has to be the same, regardless of the surrounding medium. So red, for example, *always* has a frequency of $4.41 \times 10^{14}$ Hz. But its wavelength in any medium, $\lambda = c/4.41 \times 10^{14}$, depends on the speed $c$ of red light in the medium. In fact.

the wavelength of red light in air is 680 nm ($c = 299{,}792{,}458$ m/s);

the wavelength of red light in water is 511 nm ($c = 225{,}410{,}000$ m/s);

the wavelength of red light in glass is 425 nm ($c = 187{,}370{,}000$ m/s).

Although it is the *frequency* of any color that is invariant, it is common industrial (and artistic) practice to refer to a color by wavelength. And since wavelength depends on the medium, a specific medium must be presumed. (Of course when light enters the eye, its wavelength adapts to the lightspeed in the eye's fluid.[2,3]).

## 4.6 Frequency Analysis

If you're strong in math, you probably smiled when you read "white light was made up of a superposition of colored light rays … that … were sine waves." While serving as scientific adviser to Napoleon's army, Jean-Baptiste Joseph Fourier (1768–1830) proposed that virtually *every* mathematical function was a superposition of sine waves, and the details of this *Fourier analysis* pervade every technical discipline.

We require a few tidbits from Fourier's theory to discuss color. First of all, since the set of all frequencies in the visible spectrum is a continuum, the "superposition" is an integral, not a discrete sum. In fact, there are no pure sine waves in the real world; true sinusoids endure forever, with no starting time nor ending time. Mathematically, a wave of finite duration is always a superposition (integral) of an infinite number of sinusoids of different frequencies:

$$V(t) = \int_0^\infty E(f)\,\cos[ft + \varphi(f)]\,df.$$

[2] (Is that the vitreous humor?).

[3] (No. I'm quite serious.).

We speak loosely when we say "monochrome" or "sine wave" in this context, understanding that the frequencies contributing to the integral are clustered around some dominant one. An *ideal* monochrome would need a delta function to describe its frequency density (Fig. 4.23):

By the same token, we would not necessarily classify light as "white" simply because it was a superposition over all visible frequencies. Ideal white light contains all (visible) frequencies *in equal amounts.* But if all frequencies were present in *significant but not necessarily equal* amounts, we'd possibly still classify it as white (Fig. 4.24).

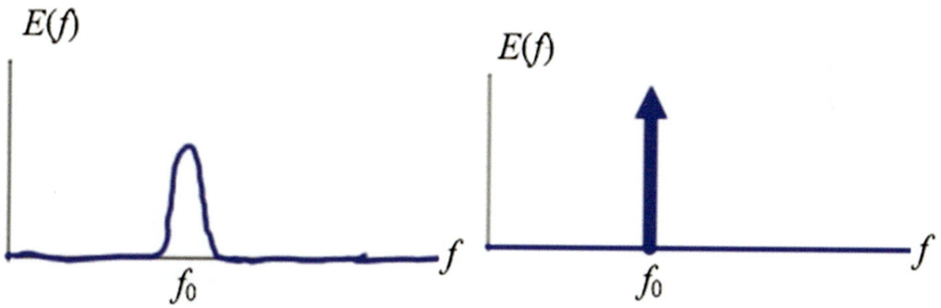


**Fig. 4.23** Physical monochrome; ideal monochrome

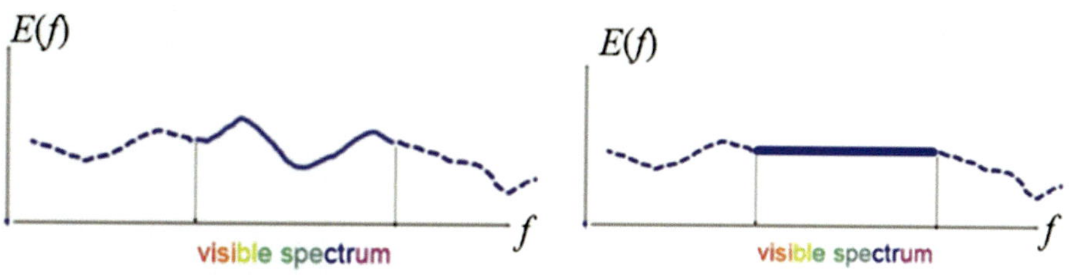


**Fig. 4.24** Physical white light; ideal white light

When a particular narrow frequency interval is dominant in the superposition, we call the light by the corresponding color, using words like "saturation" or "chroma" to describe its degree of purity. (We shall elaborate on the notions in Chap. 5.) We are similarly casual when we refer to light with a broad, but not necessarily uniform, spectrum as "white."

One last point. Everyday objects do not *generate* any colors; they can only reflect colors that shine on them. In fact, they *absorb* (or transmit, if they are transparent) all the colors except what's reflected. A blue object becomes invisible if it is illuminated with red or green light, ideally; its *blue* appearance has to be provided by the external light source. Look at the scenic photograph in Fig. 4.25. When the original is illuminated by red, green, and blue light, the features that contain none of the illuminating color are practically blacked out. Colored objects don't intrinsically have color; it has to be provided by a source. Everyone is familiar with the color of, say, cherries; they *absorb* most colors except red, and grapes absorb most except blue.

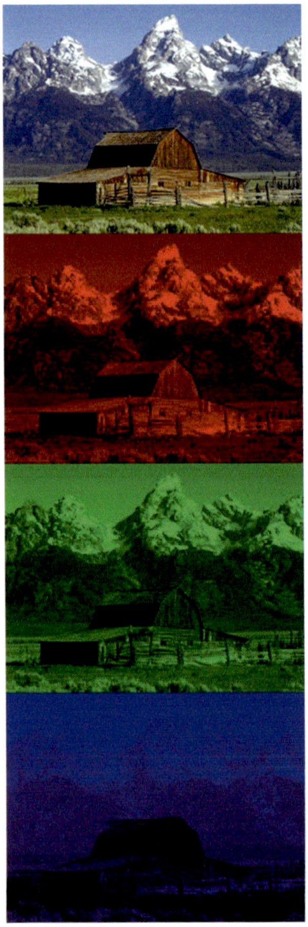

**Fig. 4.25** White, red, green, blue illumination—(The original uploader was Mike1024 at English Wikipedia.—Transferred from en.wikipedia to commons.) Transfer was stated to be made by User:nopira., public domain, https://commons.wikimedia.org/w/index.php?curid=3203703

We're not so familiar with light *sources*, which truly *emit* light:

1. Electronic transitions in atoms produce monochromatic light (with narrow frequency bands) (Figs. 4.26 and 4.27).

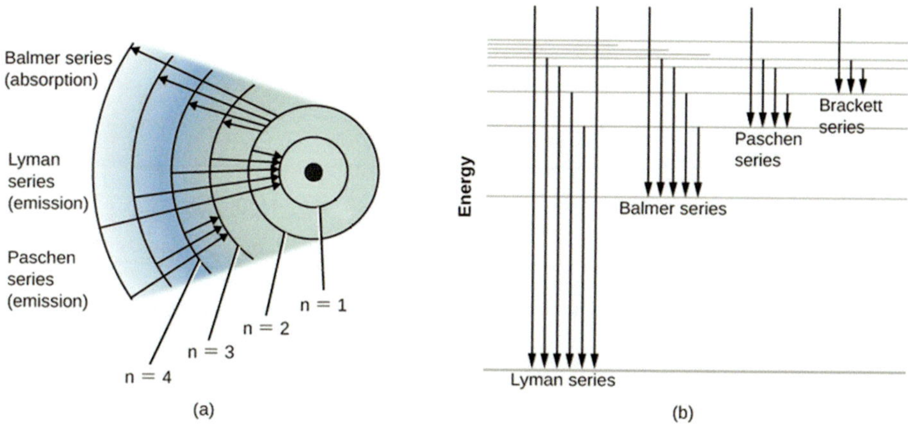


**Fig. 4.26** Hydrogen electronic transitions—By Ivan Ramirez—OpenStax astronomy, CC BY 4.0, https://commons.wikimedia.org/w/index.php?curid=58268694

**Fig. 4.27** The Balmer series produces monochromatic waves in the visible spectrum—By Merikanto, Adrignola—File:Emission spectrum-H.png, CC0, https://commons.wikimedia.org/w/index.php?curid=16417920. https://en.wikipedia.org/wiki/Hydrogen_spectral_series. Cropped and enhanced from original

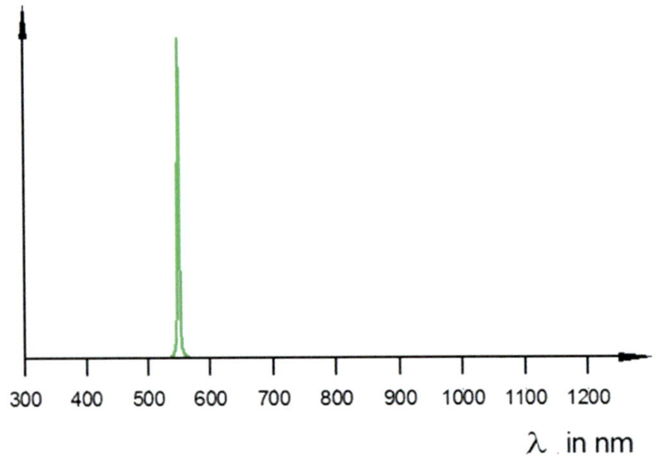


**Fig. 4.28** Green laser—By bautsch—Own work, CC0, https://commons.wikimedia.org/w/index.php?curid=42966918

2. Among the more common light sources, lasers are monochromatic (Fig. 4.28), while sunlight and fluorescent light (Fig. 4.29) nearly achieve the uniform frequency distribution we seek for white light (Fig. 4.24)—the fluorescent spectrum is spiky, but it manages to supply a diverse assortment of colors. Spectral characteristics of other sources are displayed at https://micro.magnet.fsu.edu/primer/lightandcolor/lightsourcesintro.html. (Note that incandesent tungsten bulbs are white with a slight orangish tinge, and white LED lights are noticeably bluish. Mercury lamps have several frequency clusters distributed throughtout the spectrum and thus are employed for white light, but there is a predominant bluish component to them—as is well known to night drivers).[4]

[4] If you are (wisely) reluctant to shine a light source directly into your eye, you can observe its colors by shining it onto a white object.

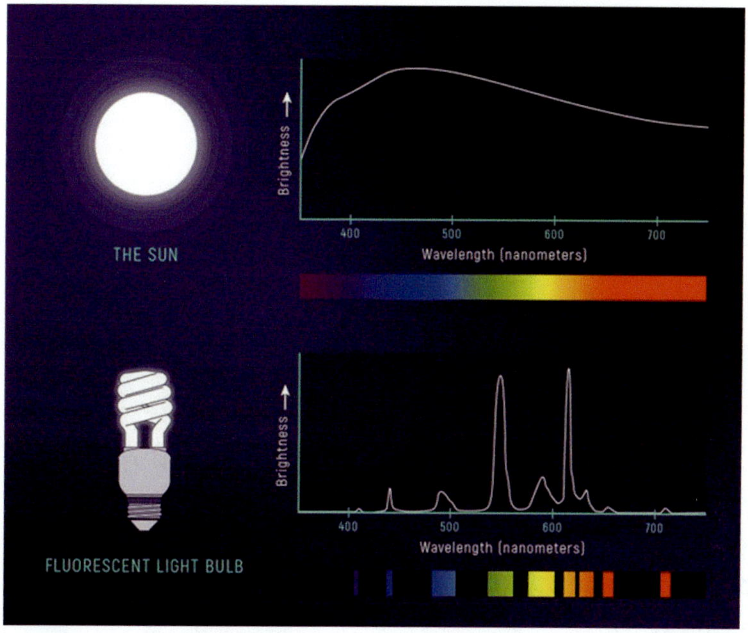


**Fig. 4.29** Sunlight and fluorescent lightbulb spectra—NASA, European Space Agency, and L. Hustak (Space Telescope Science Institute). Public domain. https://webbtelescope.org/contents/articles/spectroscopy-101--introduction

## 4.7 The Eye

If Providence had asked me to design the eye's color sensor, the first thing that would have occurred to me would be a device that responds to light with a voltage proportional to the frequency of the light; the brain would sense the voltage and identify the color. The *spectrum* of visible wavelengths runs from about $4 \times 10^{14}$ to $8 \times 10^{14}$ Hz, so its response curve would look like Fig. 4.30.

But this won't work. What if light from a green source and a blue source entered the eye simultaneously? If the sensor were linear it would add the voltages and the brain would think it's receiving red light.[5,6]

Having been rebuffed, I would next suggest an *array* of different sensors. Not one for each frequency, of course, since there is a continuum of frequencies; but each could respond to a *band* of frequencies. The intensity of the response would be proportional to the amount of light, in that frequency band, entering the sensor. Figure 4.31 depicts a crude 4-sensor array.

---

[5] (That's a source of another color.).

[6] (Oh come on, like you weren't thinking the same thing!).

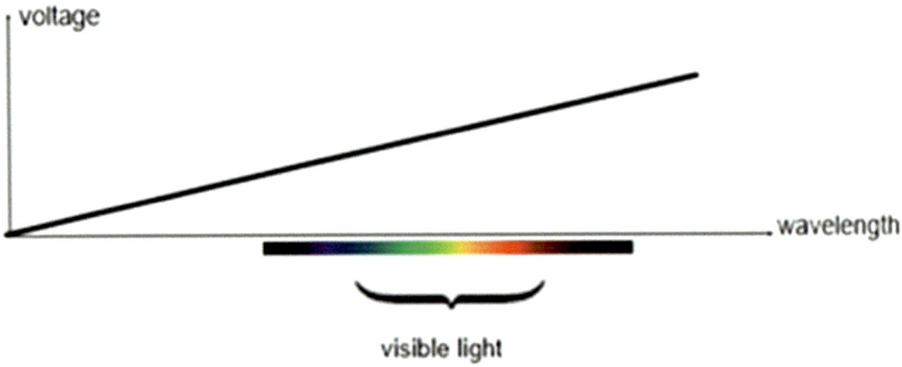


**Fig. 4.30** Proposed color sensor—Image by Johannes Ahlmann. CC BY 2.0 DEED. https://www.flickr.com/photos/entirelysubjective/6146868258. Cropped and embellished from original

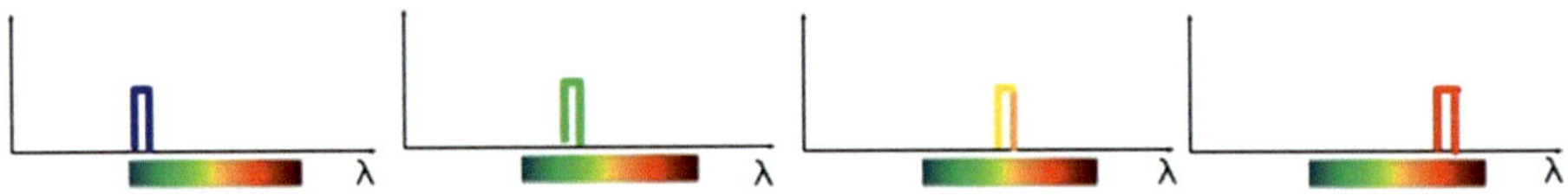


**Fig. 4.31** Eye sensor array—images by Johannes Ahlmann. CC BY 2.0 DEED. https://www.flickr.com/photos/entirelysubjective/6146868258. Cropped and embellished from original

But with (say) 1,000 sensors, the brain could discern frequencies to an accuracy of

$$\frac{8 \times 10^{14} - 4 \times 10^{14}}{10^3} = 4 \times 10^{11}\text{Hz},$$

by noting which sensors are active.

The human eye's true color-detecting mechanism works something like my second proposal. The German biophysicist/physiologist Hermann L. F. von Helmholtz (1821–1894) postulated that the human eye contains three types of color sensors, called *cones*. Their frequency responses overlap (Fig. 4.32).

The response functions $\Psi(\lambda)$ plotted in the figure are normalized so that they each have a maximum of one. *Blue cones* have their maximum in the blue region of the spectrum at a wavelength of about 420 nm. *Green cones*'s maximum, at about 530 nm, lies in the green region, and *red cones*'s maximum occurs in the red–orange region at about 560 nm.[7] The designation *S*, *M*, *L* is also used, flagging the cones with maxima at the *s*hortest, *m*iddle, and *l*ongest wavelength (Fig. 4.32).

[7] https://www.sciencedirect.com/topics/engineering/peak-sensitivity (H. F. O. Muller, *Sustainability, Energy, and Architecture* 2013s 9.4).

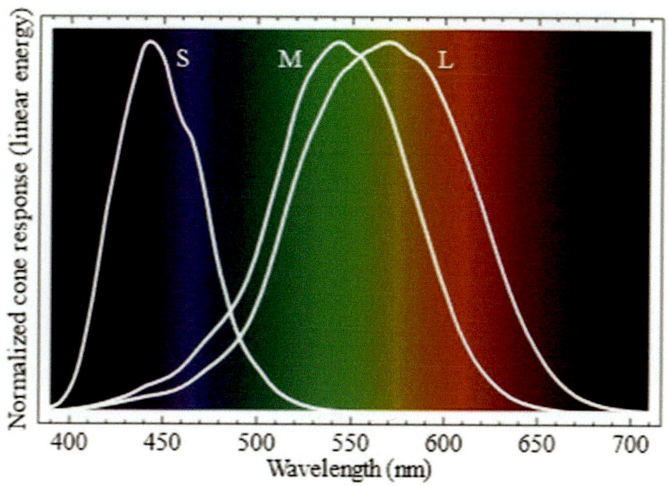


**Fig. 4.32** Cone frequency responses—By BenRG—Own work, Public domain, https://commons.wikimedia.org/w/index.php?curid=7873848. Based on the Stockman and Sharpe (2000) 2° cone fundamentals http://www.cvrl.org/cones.htm

The way we interpret these curves is as follows (Fig. 4.33). If (green) light at a wavelength of 530 nm illuminates a green cone, it transmits its maximum response $\Psi_{green}(530) = 1$ to the brain. If (blue) light at 460 nm illuminates the green cone, the response is about $\Psi_{green}(460) = 0.1$ or 10% of the maximum. Incoming light with a total intensity of $I$ candelas (1 cd = 1/683 Watts per steradian) distributed over wavelength with a *spectral density* $i_\Lambda(\lambda)$ (cd/nm),

$$I = \int_0^\infty i_\Lambda(\lambda)\; d\lambda \text{ (cd)},$$

would induce a voltage in the green cone of

$$V_{green} = \int_0^\infty \Gamma_g \Psi_{green}(\lambda)\; i_\Lambda(\lambda)\; d\lambda\text{(volts)}, \tag{4.1}$$

where $\Gamma_g$ (V/cd) is called the *gain* of the green cone.[8]

[8] $\Gamma_g$ Is the *voltage* produced by an intensity "spike" of 1 cd concentrated at the wavelength $\lambda = 530$ nm where the green cone is most sensitive ($\Psi_{green}(530) = 1$).

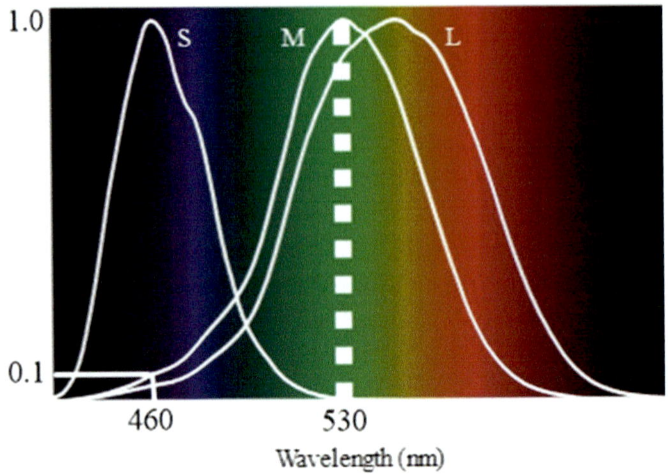


**Fig. 4.33** Contributions to $V_{green}$ of a spike of blue light at wavelength 460 nm (Embellishment of Fig. 4.32)

Notice that the $\Psi_{green}(\lambda)$ response levels triggered by the blue light at 460nm and the red light at 625 nm are the same, since $\Psi_{green}(460) = \Psi_{green}(625)$ (Fig. 4.34). Therefore the brain could not distinguish between an I candela "spike" of blue light and an I candela spike of red light from the voltage information delivered by the green cones. Fortunately the blue and the red cones can distinguish between the two monochromes.

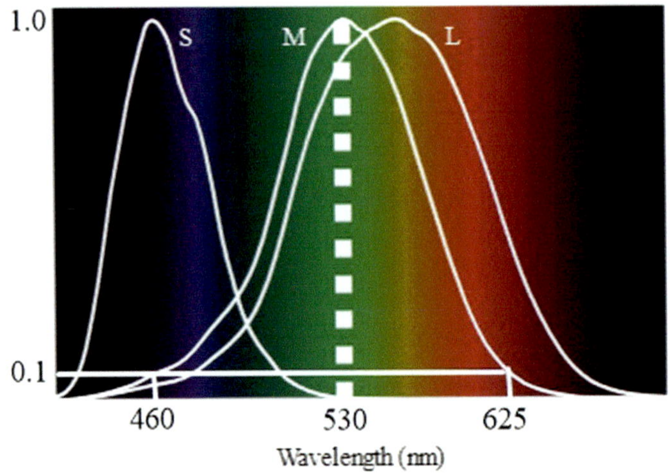


**Fig. 4.34** $\Psi_{green}(460) = \Psi_{green}(625)$ (Embellishment of Fig. 5.32)

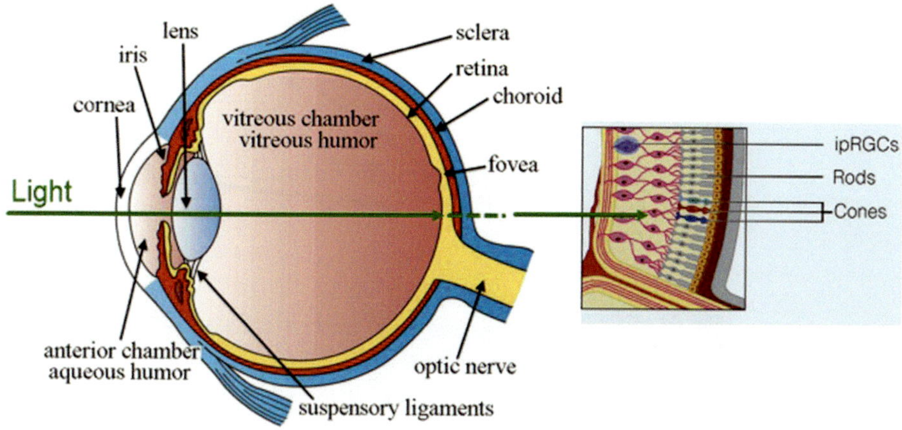


**Fig. 4.35** Cones and rods in a layer of the retina *Left*: By Holly Fischer—http://open.umich.edu/education/med/resources/second-look-series/materials—Eye Slide 3, CC BY 3.0, https://commons.wikimedia.org/w/index.php?curid=24367145. *Right*: By Christine Blume, Corrado Garbazza & Manuel Spitschan—Blume, C., Garbazza, C. & Spitschan, M. Effects of light on human circadian rhythms, sleep and mood. Somnologie 23, 147–156 (2019). https://doi.org/10.1007/s11818-019–00,215-x, CC BY 4.0, https://commons.wikimedia.org/w/index.php?curid=97114926. Cropped and embellished from originals

*Rods* are another type of light-detecting cell, but they do not contribute significant color information. In fact rods are more sensitive to light than cones, and are responsible for night vision. The parts of the eye are depicted in Fig. 4.35.

EXERCISE. The color-sensitive cones are highly concentrated at the back of the eyeball, while the color-immune rods are off to the sides. An experiment demonstrates this convincingly. Have a friend select a crayon without revealing the color to you, and hold it behind your right ear where you cannot see it—about one foot from your head. While you cover your left eye and focus your vision straight ahead, have him/her bring the crayon slowly around your head, to the front (Fig. 4.36). There will be a point, a little in front of your ear, where you will be able to see the crayon, but you will not be able to discern its color—until it is positioned more to the front. That is because the light rays from the crayon entering the eye through the lens illuminate the area populated by the color-insensitive rods first, then strike the cones as the crayon is rotated to the front.

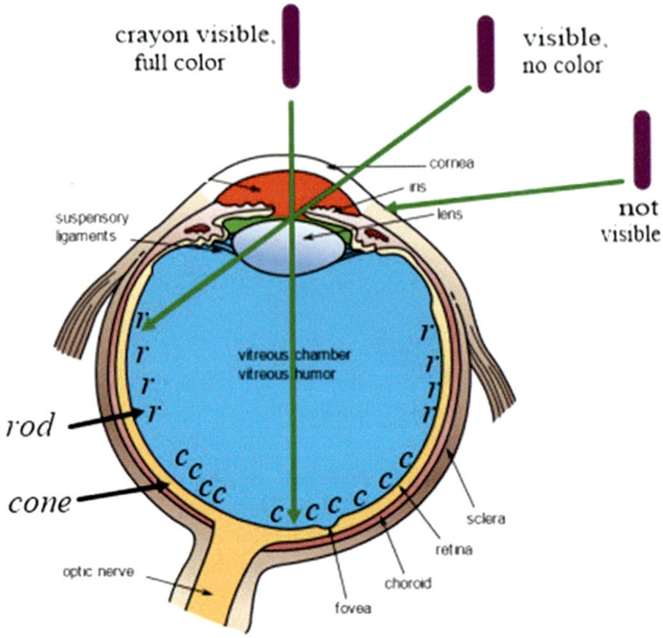


**Fig. 4.36** Right eye: Cone and rod exercise—By original: Holly Fischer vector: Pixelsquid—own work based on: Three internal chambers of the eye.png: by Holly Fischer, CC BY 3.0, https://commons.wikimedia.org/w/index.php?curid=99491006 Cropped and embellished from original

EXERCISE. There is a "blind spot"—no cones nor rods—near the back of eyeball where the optic nerve attaches. Light that enters the eye and strikes the blind spot will not be detected. To verify this, close your left eye and hold your index fingers next to each other, in front of your right eye, one foot away from your head. As you focus on your left fingernail, move your right hand to the right (Fig. 4.37). At some point you will not be able to see your right fingernail; but it will reappear when you move your hand further.

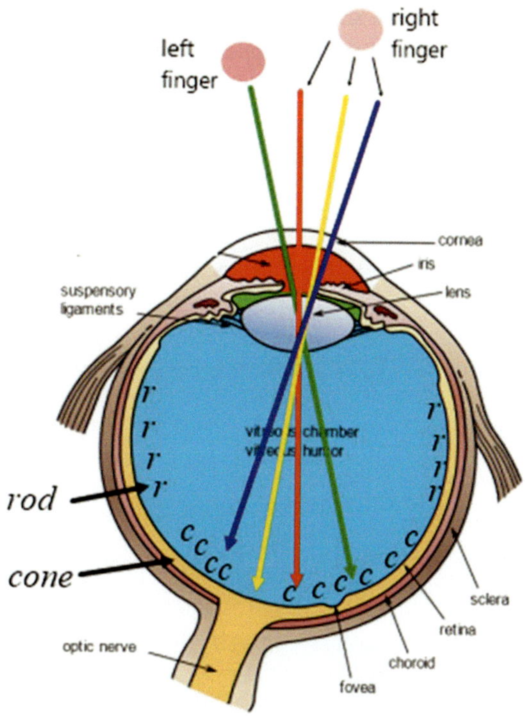


**Fig. 4.37** Right eye: Blind spot exercise—By original: Holly Fischer vector: Pixelsquid—own work based on: Three internal chambers of the eye.png: by Holly Fischer, CC BY 3.0, https://commons.wikimedia.org/w/index.php?curid=99491006 Cropped and embellished from original

## 4.8 Additive Color Phenomena

Figure 4.38 illustrates a phenomenon well known to theatrical lighting technicians. Techies have filters that enable them to shine lights of selected colors, but they can extend the range of colors by superposing the filtered lights. The theory embraced by

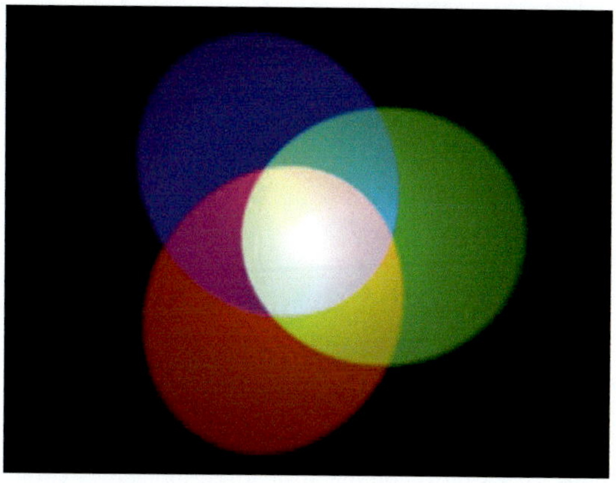

**Fig. 4.38** Stage lighting—By Pko—own work, Public domain, https://commons.wikimedia.org/w/index.php?curid=1180142

formula 4.1 of Sect. 4.7 can be used to explain this; we'll show why a superposition of green and red light will appear yellow.

See Fig. 4.39. Suppose we shine 1 cd of 560-nm (green) light and 1 cd of 630-nm (red) light onto a screen—and view it. The formula says that a green cone would experience a voltage excitation of (Fig. 4.40).

$$V_{green} = \int_0^\infty \Gamma_g \Psi_g(\lambda)\ i_\Lambda(\lambda)\ d\lambda$$

$$= \Gamma_g[\Psi_g(560) \times 1 + \Psi_g(630) \times 1] \approx \Gamma_g[.89 + .08] = .97\Gamma_g \text{volts}. \quad (4.2)$$

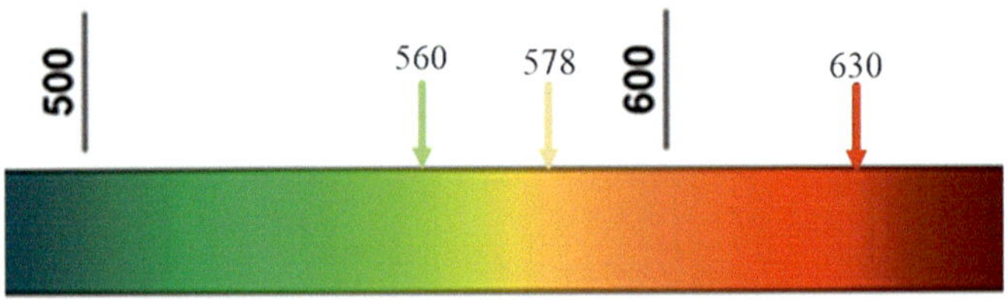


**Fig. 4.39** Color addition example—By Johannes Ahlmann. CC BY 2.0 DEED. https://www.flickr.com/photos/entirelysubjective/6146868258. Cropped from original

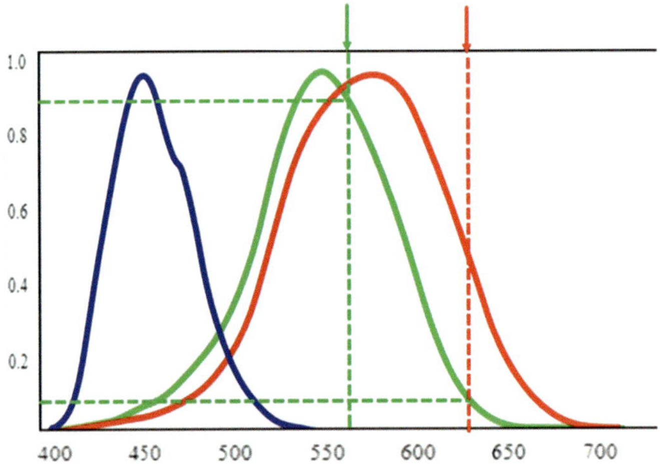


**Fig. 4.40** Green cone response to green and red light—by Johannes Ahlmann. CC BY 2.0 Deed https://www.flickr.com/photos/entirelysubjective/ Cropped and embellished from original

And a red cone experiences (Fig. 4.41)

$$V_{red} = \Gamma_r[\Psi_r(560) \times 1 + \Psi_r(630) \times 1] \approx \Gamma_r[.95 + .48] = 1.43\Gamma_r \text{volts}.$$

A blue cone receives practically 0 volts since $\Psi_b(560) \approx \Psi_b(630) \approx 0 \Rightarrow V_{blue} \approx 0$ (Fig. 4.42).

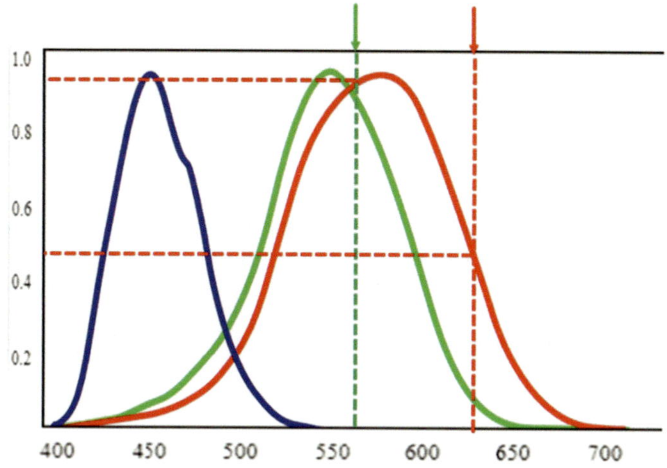


**Fig. 4.41** Red cone response to green and red ligh—by Johannes Ahlmann. CC BY 2.0 Deed https://www.flickr.com/photos/entirelysubjective/ Cropped and embellished from original

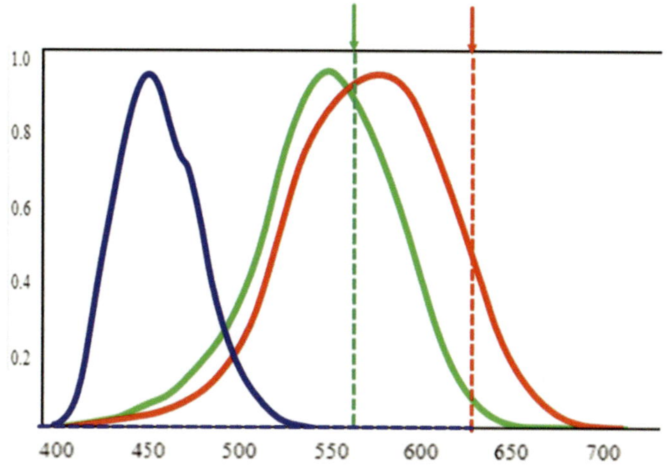


**Fig. 4.42** Blue response to green and red light—by Johannes Ahlmann. CC BY 2.0 Deed https://www.flickr.com/photos/entirelysubjective/ Cropped and embellished from original

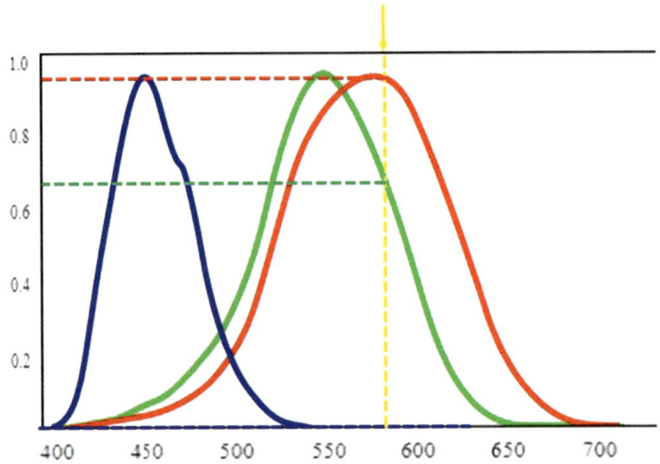


**Fig. 4.43** Cone responses to yellow light—by Johannes Ahlmann. CC BY 2.0 Deed https://www.flickr.com/photos/entirelysubjective/ Cropped and embellished from original

But what cone voltages would 1.5 cd of 578 nm (yellow) light generate? Consulting Fig. 4.43, we again see $V_{\boldsymbol{blue}} \approx 0$, and

$$V_{\boldsymbol{green}} = \Gamma_g \Psi_g(570) \times 1.5 \approx \Gamma_g[.67 \times 1.5] \approx 1.005\Gamma_g \text{volts},$$

$$V_{\boldsymbol{red}} = \Gamma_r \Psi_r(570) \times 1.5 \approx \Gamma_r[.95 \times 1.5] \approx 1.425\Gamma_g \text{volts}.$$

That is,

$$\begin{bmatrix} V_{green} \\ V_{red} \\ V_{blue} \end{bmatrix}_{green+red} \approx \begin{bmatrix} 0.97 \\ 1.43 \\ 0 \end{bmatrix}; \begin{bmatrix} V_{green} \\ V_{red} \\ V_{blue} \end{bmatrix}_{yellow} \approx \begin{bmatrix} 1.005 \\ 1.425 \\ 0 \end{bmatrix}.$$

The (voltage) message sent to the brain by the cones exposed to the green-plus-red combination is practically the same as that sent by those receiving the monochrome yellow light— "green light plus red light equals yellow light", as illustrated in Fig. 4.38. *Keep in mind that if the eye had a larger variety of cones, it couldn't be "fooled" in this manner.*

EXERCISE. Figure 4.38 also indicates blue light plus green light produces cyan light. Show that 1 cd of 480-nm blue light superposed on 1.2 cd of 530 nm green light gives cone responses nearly identical to those of 3.2 cd of 495-nm cyan light (see Fig. 4.44).

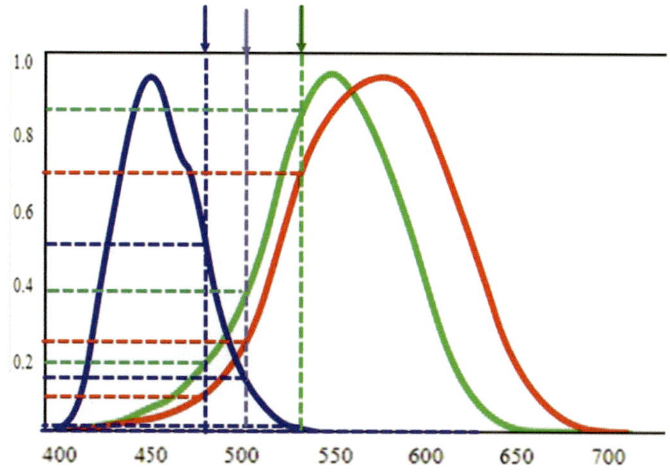


**Fig. 4.44** Cone responses to blue, cyan, and green light—by Johannes Ahlmann. CC BY 2.0 Deed https://www.flickr.com/photos/entirelysubjective/ Cropped and embellished from original

Interestingly, addition of blue light to red light produces magenta (Fig. 4.38), a color that does not appear in the prism/rainbow spectrum (Fig. 4.10). Such colors are called *non-spectral* colors; other examples are brown, silver, plum, and gold. Colors like yellow and cyan which *do* lie in the spectrum but whose cone responses can also be synthesized by addition of spectral colors are called *metamers.*

What is the significance of the white spot in the threefold overlap in Fig. 4.38? White light has a uniform spectrum (Fig. 4.24), while the monochrome spectrum consists of three spikes; we can't synthesize true white light with three monochromes.

**BUT**… We can *fool* the brain into *thinking it sees white* if we can tune the three monochromes so that they reproduce the *same three cone voltage values* that white light produces! (After all, we just used green and red to fool the cones into thinking they were seeing yellow.) We only need to deceive three sensors, and we have three controls at our disposal. We'll show that we can "*synthesize*" white light using 465 nm blue, 520 nm green, and 615 nm red light.

First we calculate the voltages generated in the blue, green, and red cones generated by *true* white light, with unit spectral density, $i_\Lambda(\lambda) \equiv 1$ cd/nm. From formula 4.1 (Sect. 4.7),

$$V_{blue} = \int_0^\infty \Gamma_b \Psi_b(\lambda)\ i_\Lambda(\lambda)\ d\lambda = \Gamma_b \int_0^\infty \Psi_b(\lambda) d\lambda,$$

$$V_{green} = \cdots = \Gamma_g \int_0^\infty \Psi_g(\lambda)\ d\lambda,$$

$$V_{red} = \cdots = \Gamma_r \int_0^\infty \Psi_r(\lambda)\, d\lambda.$$

The integrals of the three normalized frequency response curves are numerically estimated from Fig. 4.32 to be.

$$\int_0^\infty \Psi_b(\lambda)\, d\lambda \approx 56, \int_0^\infty \Psi_g(\lambda)\, d\lambda \approx 100, \int_0^\infty \Psi_r(\lambda)\, d\lambda \approx 116,$$

So the cone voltages produced by this true white light are

$$V_{blue} \approx 56\Gamma_b,\ V_{green} \approx 100\Gamma_g,\ V_{red} \approx 116\Gamma_r\text{volts}.$$

Now let's calculate the cone voltages produced by three equal monochrome spikes of 83 cd each at wavelengths 465, 520, and 615 nm. Mimicking our computation in formula 4.2 above and consulting Fig. 4.31, we find

$$V_{blue} = \Gamma_b[\Psi_b(465) + \Psi_b(520) + \Psi_b(615)] \times 83 \approx \Gamma_b[.62 + .03 + 0]83 \approx 54\,\Gamma_b$$

$$V_{green} = \Gamma_g[\Psi_g(465) + \Psi_g(520) + \Psi_g(615)] \times 83 \approx \Gamma_g[.18 + .90 + .13]83 \approx 100\ \Gamma_g$$

$$V_{red} = \Gamma_r[\Psi_r(465) + \Psi_r(520) + \Psi_r(615)] \times 83 \approx \Gamma_r[.10 + .70 + .61]83 \approx 117\,\Gamma_r\text{volts}.$$

The triple-monochrome cone voltages match each of the true-white cone voltages quite well.[9]

Two comments:

1. Bees and butterflies have 4 types of cones in their retinas; they can probably sense right away if incandescent lamps have been replaced by LED lighting.
2. Dogs and cats have only 2 types of retina cells (blue and yellow). They're not truly colorblind, but they can't distinguish as many colors as humans.

EXERCISE. If you can get hold of a *light box*, try out the color addition rules mentioned in this section.

[9] The methodology for selecting the parameter values 465 nm, 520 nm, 615 nm, and 83 cd for this example was largely governed by trial and error.

## 4.9 Trichromatic Color Theory

You have seen that the addition of spectral green to spectral red produces a response indistinguishable (in humans) from that of spectral yellow, and the addition of intensities of spectral blue plus spectral red produces (nonspectral) magenta, and spectral blue plus spectral green plus spectral red produces white. So you're dying to ask: what is the *totality* of human-perceived colors that can be produced by combining various amounts of red, blue, and green light?

James Clerk Maxwell (1835–1879), following the discoveries of Thomas Young (1773 – 1829), explored this experimentally. Maxwell didn't have access to many adjustable light sources, and his method of "adding" colors was to paint them in proportions on a disk and spin the disk rapidly, so that the eye averaged out the colors (Fig. 4.45).

Let's approach this question from the engineer's viewpoint. (We'll reevaluate it from the artist's perspective later.) We want to tabulate which color we see if we add $R$ units of red light (of a specific wavelength), $G$ units of green light, and $B$ units of blue light. Conceptually, we could display the resulting colors in a three-dimensional $RGB$ space (specifically, the first octant of the space since none of the $RGB$ coordinates should be negative). We depict this with Figs. 4.46 and 4.47.

Now of course if we doubled the $RGB$ coordinates of any point, we would increase the intensity of the light, but not its color; color depends only on the *proportions* $R$:$G$:$B$. Every ray from the origin in RGB space contains the same color; only the intensity increases along the ray.

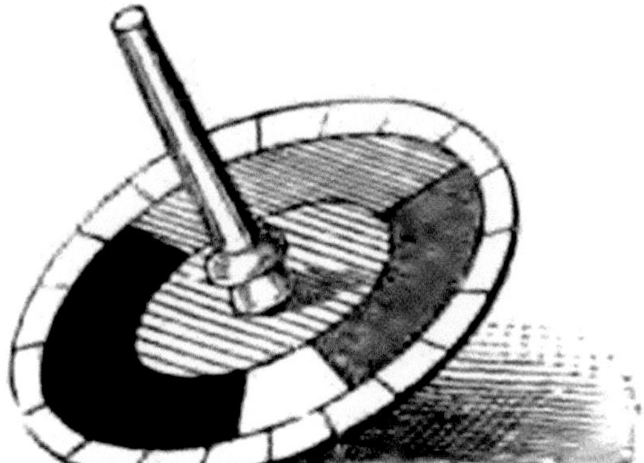

**Fig. 4.45** Maxwell color top—By Edward wheeler scripture—Thinking, feeling, doing, public domain, https://commons.wikimedia.org/w/index.php?curid=3334687. Cropped and embellished from the original

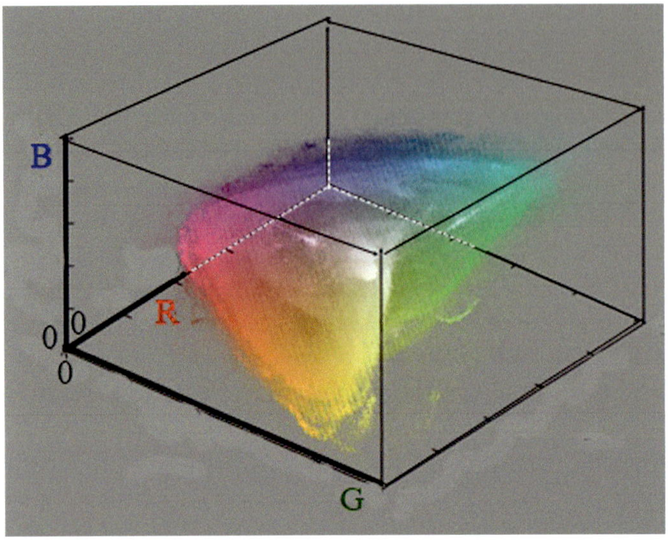


**Fig. 4.46** First octant of RGB space—By Niabot—Own work, CC0, https://commons.wikimedia.org/w/index.php?curid=12407052 Cropped and embellished from the original

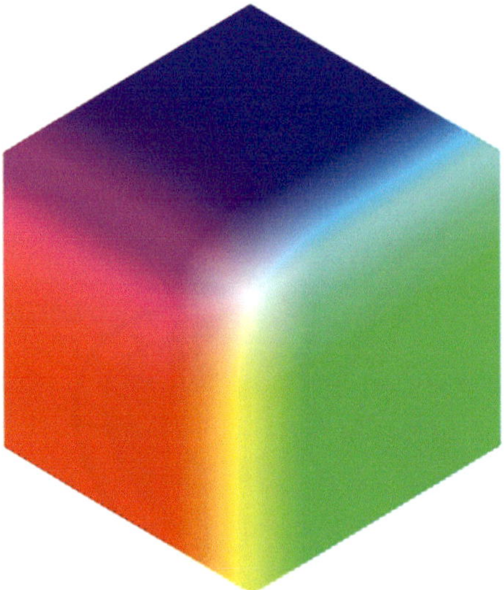

**Fig. 4.47** RGB color cube—By 4throck—own work, CC0, https://commons.wikimedia.org/w/index.php?curid=110691227

Therefore we can simplify the diagram by focusing on the color—ignoring the brightness - and rescaling the coordinates so that they add up to one:

$$r = \frac{R}{R+G+B}, g = \frac{G}{R+G+B}, b = \frac{B}{R+G+B}; r+g+b = 1. \tag{4.3}$$

This shrinks the octant down to a *triangle* of distinct colors in *r,g,b* space with corners (1,0,0), (0,1,0), and (0,0,1). See Fig. 4.48, where we exhibit this *Maxwell triangle.*

(To be even more efficient, we can project this display onto a right triangle in the *r,g* plane by simply omitting the third coordinate, and remembering $b = 1 - r - g$. (This is easy to remember if we think of *r, g,* and *b* as *percentages*; if $r = 20\%$ and $g = 50\%$, we know $b = 30\%$.) Note that when $r = 1$, then $g = b = 0$ and the corner is "pure" red. The blue corner falls at $r = g = 0$, etc.)

The triangle shows that a huge array of colors can be synthesized by combining red, green, and blue light. Of course the span of colors obtained depends on the choice of the shades of red, green, and blue used for the corners of the color triangle. Any particular choice defines the *primary colors* for the additive scheme. In 1931 the International Commission on Illumination (C.I.E.), guided by the experiments of William David Wright and John Guild, recommended the use of the spectral 700-nm red, 546.1-nm green, and

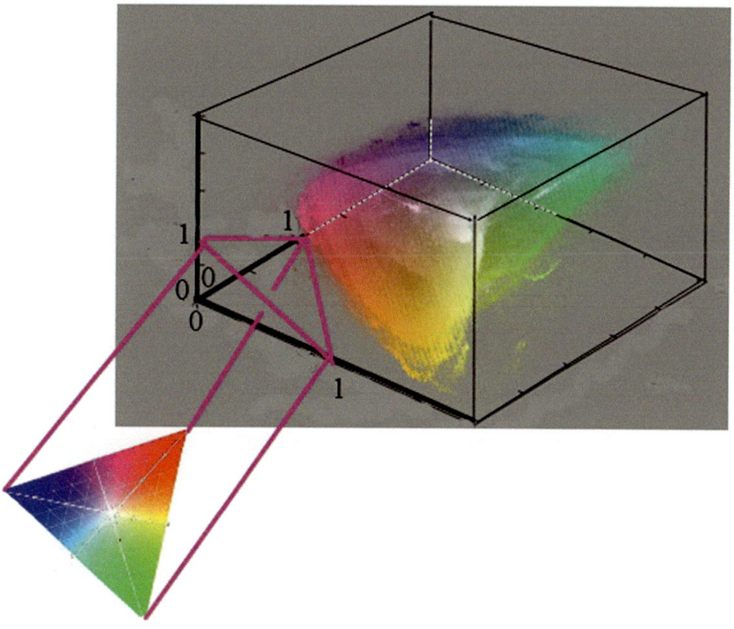


**Fig. 4.48** Maxwell color triangle—By Niabot—own work, CC0, https://commons.wikimedia.org/w/index.php?curid=12407052; and By Polina Tishina—Polina Tishina, Public domain, https://commons.wikimedia.org/w/index.php?curid=9816956. Cropped and embellished from the original

435.8-nm blue as primary colors, officially defining what is known as the *RGB color space*. Several factors went into this decision, including the robustness of the color triangle and the availability of light sources.

The web site https://en.wikipedia.org/wiki/CIE_1931_color_space discusses many issues involved with the choice of primary colors. It shows the proportions of the primaries needed to synthesize the spectral colors; that is, the "recipe" for each of the spectral colors. The recipes, using the C.I.E.'s RGB "ingredients", are depicted in the graph of Fig. 4.49. (The "ingredients" $(\overline{r}, \overline{g}, \overline{b})$ are further rescaled versions of our (*r*, *g*, *b*); we omit the details.)

Thus at $\lambda = 546.1$ nm, which is the green primary, we have $R = B = 0$. And analogous conditions pertain at 435.8 nm and 700 nm.

Right away we see in the graph that the C.I.E. recipe calls for a *negative* amount of red in the mix when the synthesized color's spectral wavelength lies in the blue-green range 440–540 nm. So C.I.E.'s choice of primaries (700-nm red, 546.1-nm green, 435.8-nm blue) is still not robust enough to span all the spectral colors.

Consequently in the same year (1931) that they defined the *RGB* color space, the C.I.E. defined *the XYZ color space* employing a matrix mapping of the RGB coordinates, given by the equation

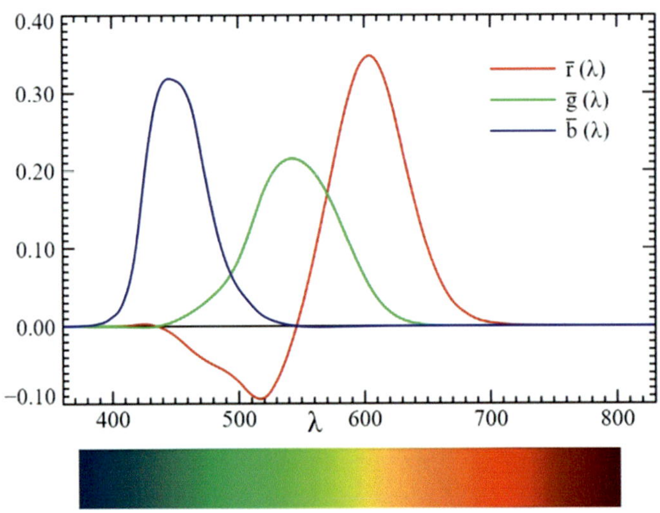


**Fig. 4.49** CIE recipes for the spectral colors—By Marco Polo at English Wikipedia—Transferred from en.wikipedia to Commons by Kanie., Public Domain, https://commons.wikimedia.org/w/index.php?curid=3659575; and by Johannes Ahlmann. CC BY 2.0 DEED. https://www.flickr.com/photos/entirelysubjective/6146868258. Cropped and embellished from original

$$\begin{bmatrix} X \\ Y \\ Z \end{bmatrix} = \begin{bmatrix} 2.7688 & 1.7517 & 1.1301 \\ 1 & 4.5906 & 0.0601 \\ 0 & 0.0565 & 5.5942 \end{bmatrix} \begin{bmatrix} \mathrm{R} \\ \mathrm{G} \\ \mathrm{B} \end{bmatrix}. \tag{4.4}$$

The corresponding *normalized XYZ* coordinates are

$$x = \frac{X}{X+Y+Z}, y = \frac{Y}{X+Y+Z}, z = 1 - x - y = \frac{Z}{X+Y+Z}. \tag{4.5}$$

After this mapping, all of the visible colors have *positive* $(x,y,z)$ coordinates (Fig. 4.50): RGB's spectral green primary ($\lambda = 546.1$) has $\mathrm{R} = \mathrm{B} = 0$. Therefore the C.I.E. XYZ coordinates of this green are given by

$$\begin{bmatrix} X \\ Y \\ Z \end{bmatrix} = \begin{bmatrix} 2.7688 & 1.7517 & 1.1301 \\ 1 & 4.5906 & 0.0601 \\ 0 & 0.0565 & 5.5942 \end{bmatrix} \begin{bmatrix} 0 \\ \mathrm{G} \\ 0 \end{bmatrix} = \begin{bmatrix} 1.7517 \\ 4.5906 \\ 0.0565 \end{bmatrix} \mathrm{G},$$

corresponding to normalized coordinates

$$\begin{bmatrix} x \\ y \\ z \end{bmatrix} = \begin{bmatrix} 0.2738 \\ 0.7174 \\ 0.0088 \end{bmatrix}.$$

In Fig. 4.50 this spectral green can be identified either by its wavelength ($\lambda = 546.1$) *or* its coordinates in the normalized XY plane, (0.2738, 0.7174).

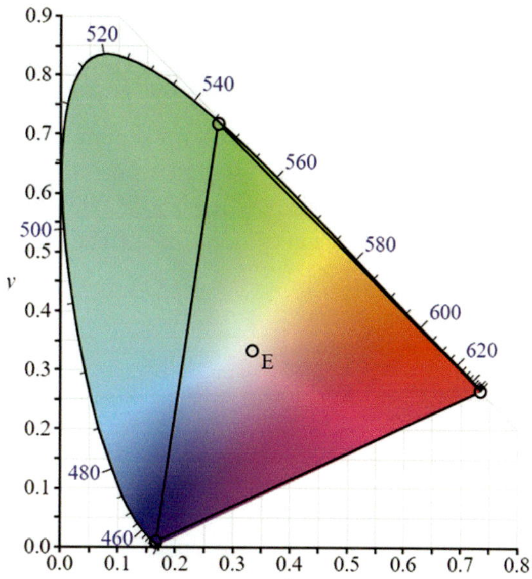


**Fig. 4.50** Normalized XYZ color coordinates for the entire visible spectrum—By BenRG—Own work, inspired by File:CIExy1931.png, Public Domain, https://commons.wikimedia.org/w/index.php?curid=7889718

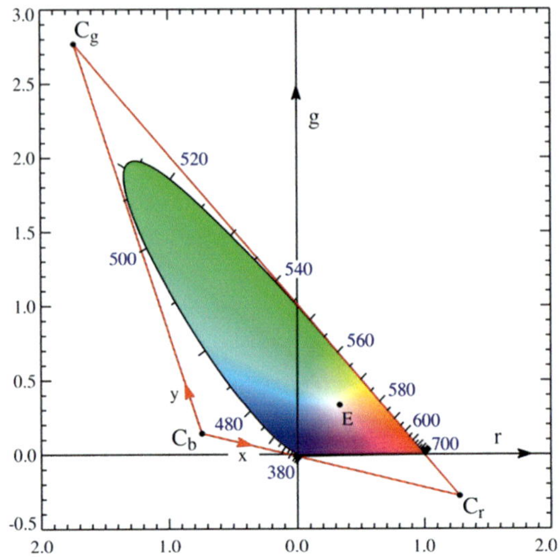


**Fig. 4.51** CIE XYZ "primary colors"—Author unknown. Public domain, https://commons.wikimedia.org/w/index.php?curid=195285

*Interesting question*: What are the (*r, g*) coordinates for the normalized XYZ *primary* colors - the points with coordinates [1,0], [0,1], and [0,0] in Fig. 4.51? As seen, they lie outside the visible spectrum. If we invert the formulas (4.4, 4.5) we find the primaries at the points shown in Fig. 4.51.

*The primaries for the* XYZ *color space are not visible to the human eye.* They are simply mathematical abstractions, invented to keep the coefficients (X, Y, Z) in the color matching functions positive.[10,11]

A final note. Of course there are limitations to how "close" two colors can be before the human eye fails to distinguish them. The website https://en.wikipedia.org/wiki/MacAdam_ellipse describes results reported by David L. MacAdam ("Visual Sensitivities to Color Differences in Daylight," D. L. MacAdam, *Journal of the Optical Society of America,* v. 32 #5. 1942, 247–274). Performing experiments on human subjects, MacAdam populated the graph in Fig. 4.50 with ellipses demonstrating the limits of the eye's spectral resolution over the regions of *XYZ* space. So beyond a certain point, pursuit of high-precision accuracy in color calculations is futile.

---

[10] They are pigments of the imagination.

[11] (My wife says I have a problem.).

# Color Theory: Technology 5

## 5.1 Subtractive Mixing

You may have noticed that in Chap. 4 I studiously avoided stating the result of combining blue and yellow. Everyone that grew up with a watercolor set knows that blue paint plus yellow paint gives green paint. But *mixing paint* is quite different from *adding light sources.*

Recall that if an object not a source, its color comes from the light that it reflects when illuminated by a white source. Yellow paint is certainly not a source of yellow light; nor is it a filter that selectively transmits yellow. It reflects the yellow component of light that shines upon it—while it absorbs all the other colors.

But this can't be the whole story. If the yellow paint absorbs all colors except yellow—and the blue paint absorbs all colors other than blue—if we combine blue paint molecules with yellow paint molecules, they absorb—what?—*all* colors? Shouldn't the mixture be black? That can't be right.

Let's assume the yellow paint's color isn't spectral yellow, but the yellow metamer described in Sect. 4.8; it's actually a combination of molecules that absorb every color but red, and molecules absorbing everything but green. By reflecting red and green it deceives the three-coned human eye into seeing it as yellow.

Now if the blue paint's color is actually the metamer cyan, obtained by combining blue reflectors and green reflectors (Fig. 4.44),[1] then the cyan paint is absorbing all colors but blue and green.

So what is the behavior of a mixture of molecules, some of which absorb all colors but red and green, and some absorbing all but blue and green? The mixture absorbs everything but green. *Metamer cyan paint mixed with metamer yellow paint reflects green.* The "blue"

[1] Creative Commons License Links for all images are listed in the Preface.



A. D. Snider, *From STEM to STEAM*, Synthesis Lectures on Engineering, Science, and Technology, https://doi.org/10.1007/978-3-031-57316-3_5

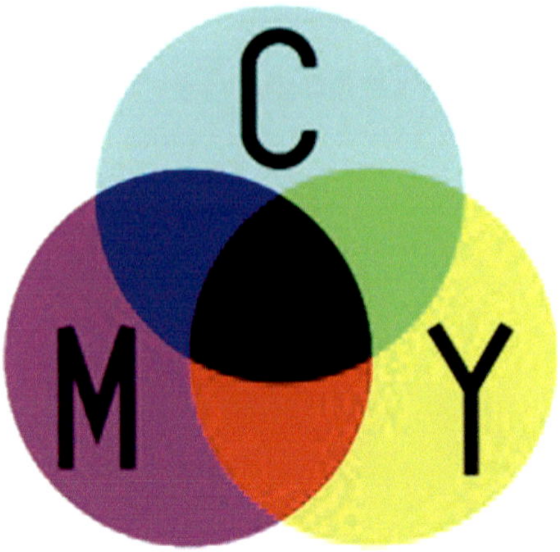


**Fig. 5.1** Subtractive color mixing—by SharkD at English Wikipedia/later version uploaded by Jacobolus, Dacium at en.wikipedia. Transferred from en.wikipedia to Commons. Public Domain, https://commons.wikimedia.org/w/index.php?curid=3791468

and "yellow" watercolors of our youthful experiments were actually metamers, and we were adding color *absorbers*. This complementary logic, where we add the colors that are absorbed—not emitted—is known as *subtractive color theory*. Just as stage lighting designers operate by the rules of additive color theory, artists (who combine paints, not sources) go by the subtractive color laws. The rules for subtractively mixing cyan, yellow, and magenta are summarized in Fig. 5.1.

**Exercise** Explain the other three absorption mechanisms illustrated in Fig. 5.1: cyan plus magenta equals blue, yellow plus magenta equals red, cyan plus yellow plus magenta equals black.

In the eighteenth century red, yellow, and blue were taken to be the primary colors for mixing pigments (paint or ink). Later, commercial printers adopted cyan, magenta, and yellow as the primaries, mainly because they provided a wider range of colors. Black was also employed, since black ink is cheaper than mixing all three primaries. Magenta is sometimes known as "process red", and cyan as "process blue." Cyan (the blue-green metamer) dye absorbs red light, its "complement". The complement of yellow (red-green metamer) is blue, and the complement of magenta (red-blue metamer) is green.

Figure 5.2 shows an example of a *color wheel*, which displays three primary colors opposite their complements, and "secondary" and "tertiary" colors generated by further mixing. The distinction between adding light sources and mixing colored dyes was first formulated in the books of J. W. von Goethe (*Theory of Colors*, 1810) and M.-E. Chevreul (*The Law of Simultaneous Color Contrast*, 1839).

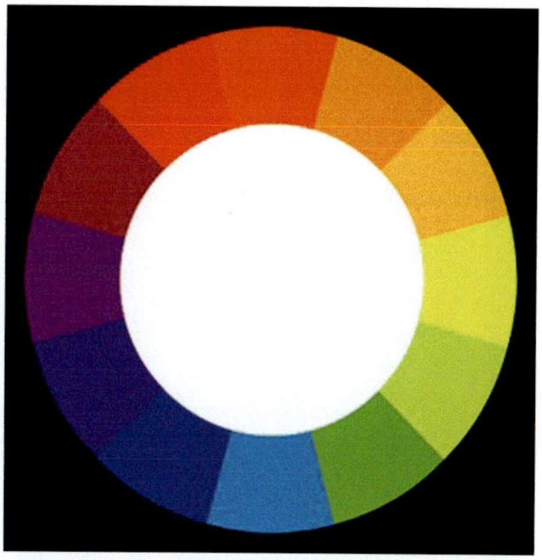

**Fig. 5.2** Artist's color wheel—by no machine-readable author provided. MarianSigler assumed (based on copyright claims). No machine-readable source provided. Own work assumed (based on copyright claims). Public Domain, https://commons.wikimedia.org/w/index.php?curid=770594

## 5.2 Color Addition by Juxtaposition

We have noted how colors from different sources can be added by shining them directly into the eye (more prudently, shining them onto a screen) or by oscillating them rapidly in the field of vision and letting the eye average them over time (Maxwell's spinning disk). Another widely-used procedure is juxtaposition—placing many tiny patches of the color sources closely together so that the eye averages them over position ("smears" them). You may be surprised how ubiquitous this practice is; take a good magnifying glass and look at a white portion of your color tv screen and your computer screen (Fig. 5.3).

The colors in these devices are all produced by adding the primary sources of red, green, and blue, although the particular arrangement differs according to the technology. If you look at a yellow image on the screen, you will see that only the red and green "pixels" have been energized. You might not even be able to discern the dark blue and green pixels if you look at a red portion.

Of course the color commands are controlled by computer logic. The relative intensities $R, G, B$ of the primaries in the addition are binary coded from 0 to 255 ($=2^8 - 1$) (Fig. 5.4).

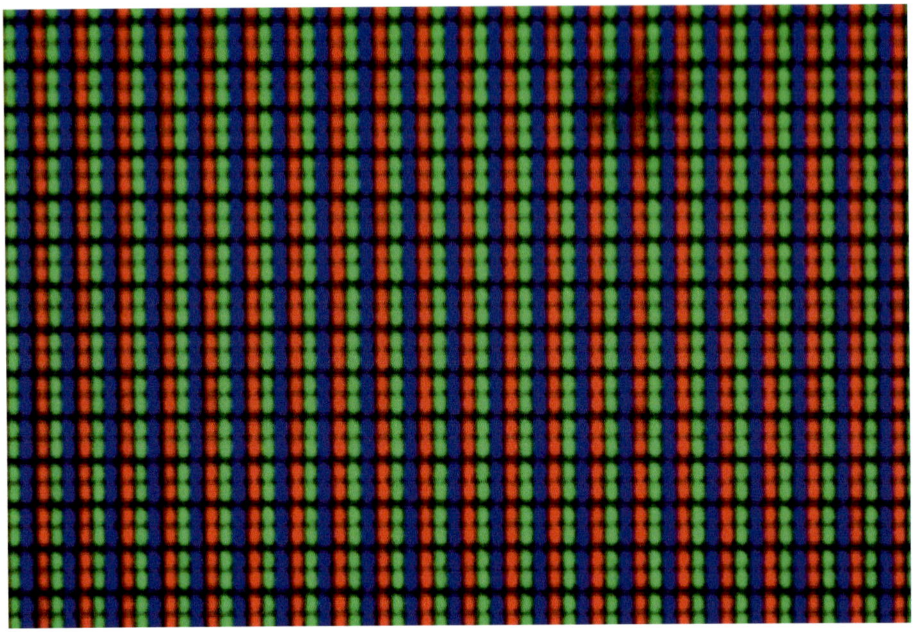

**Fig. 5.3** Close-up of screen—by Pascal. Public Domain. https://www.flickr.com/photos/pasukaru76/3635716299

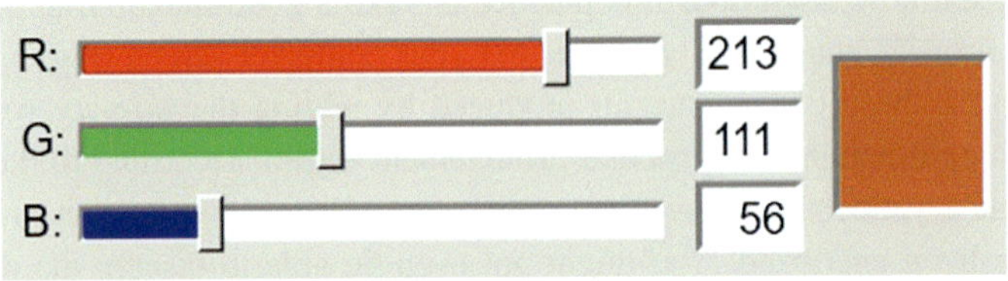


**Fig. 5.4** RGB color selector. Each slider ranges from 0 to 255. By Nicky Nouse—own work based on: RGB sliders.png, Public Domain, https://commons.wikimedia.org/w/index.php?curid=116188254

Now, apply your magnifying glass to some images from your inkjet printer. These color dots are finer, and you may need stronger magnification. In inkjet printing, colored dots that are juxtaposed combine additively, while if the dots are printed on top of each other they can combine subtractively if the inks mix (Figs. 5.5 and 5.6).

The primary colors for inkjet printing are magenta, cyan, and yellow (as you discovered the first time you bought refills). Also black ink can used for additional clarity, and to circumvent the complexity in mixing all three primaries. And by resizing the dots the white

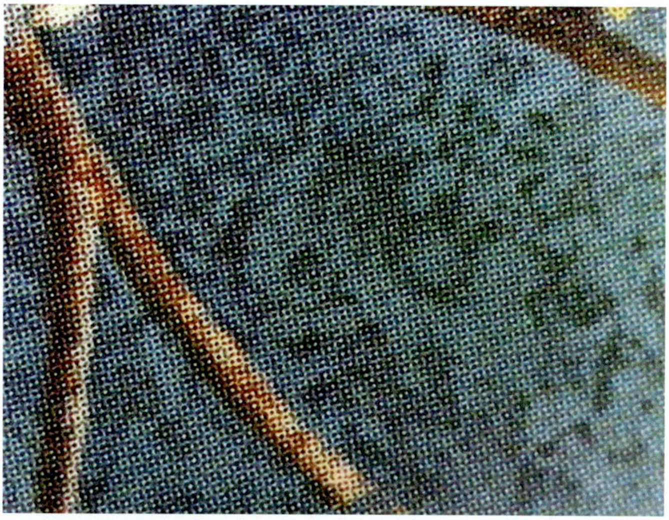

**Fig. 5.5** Printed image (magnified)—by Psiĥedelisto—own work, CC0. https://en.wikipedia.org/wiki/CMYK_color_model

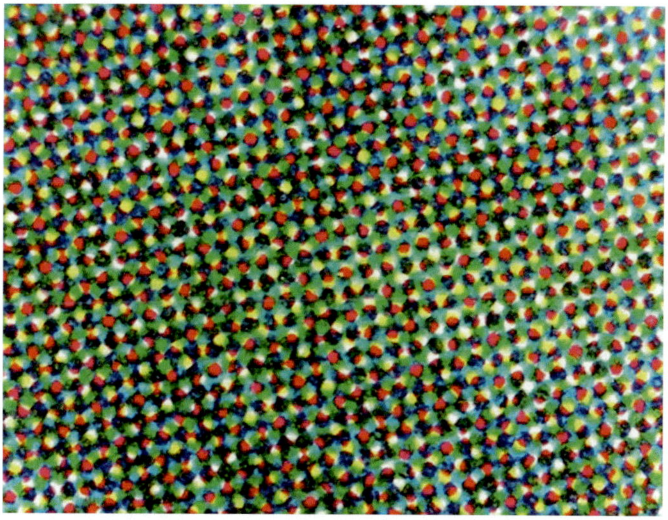

**Fig. 5.6** Enlargement of Fig. 5.5—by Psiĥedelisto—own work, CC0, https://commons.wikimedia.org/w/index.php?curid=66759793

background provided by the paper can be exploited as, effectively, another primary—a procedure known as halftoning (Fig. 5.7).

Finally: look closely at your comic books and Sunday funnies (Fig. 5.8). You probably won't need the magnifying glass to see the splotches; the quality of newspaper does not warrant the effort for fine detail.

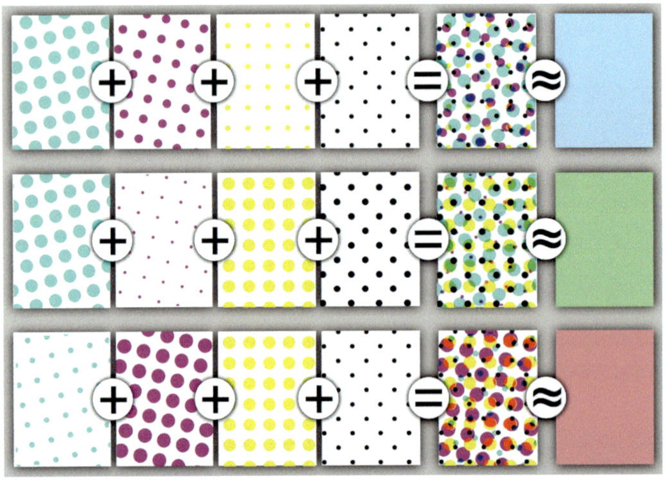

**Fig. 5.7** Larger dots ⇒ tighter juxtaposition ⇒ alteration of the perceived color—by original: Slippens Vector: Pbroks13—own work based on: Halftoningcolor.png, Public Domain, https://commons.wikimedia.org/w/index.php?curid=5656963

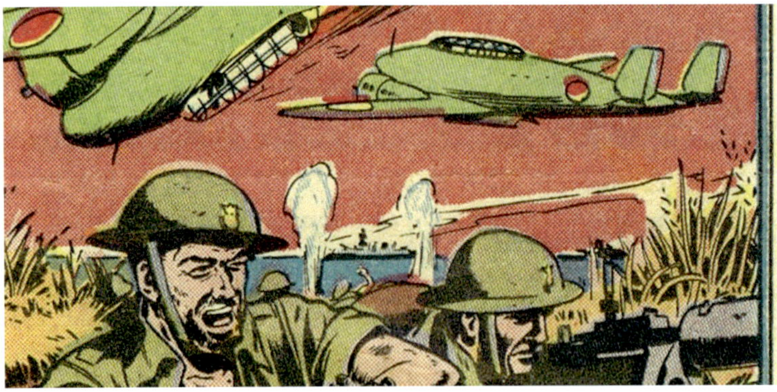

**Fig. 5.8** Secret war news military comic, March 1942. From the Herbert Freuler Collection (COLL/5550) at the Archives Branch, Marine Corps History Division. CC BY 2.0 DEED https://www.flickr.com/photos/usmcarchives/32349172066/in/photostream/

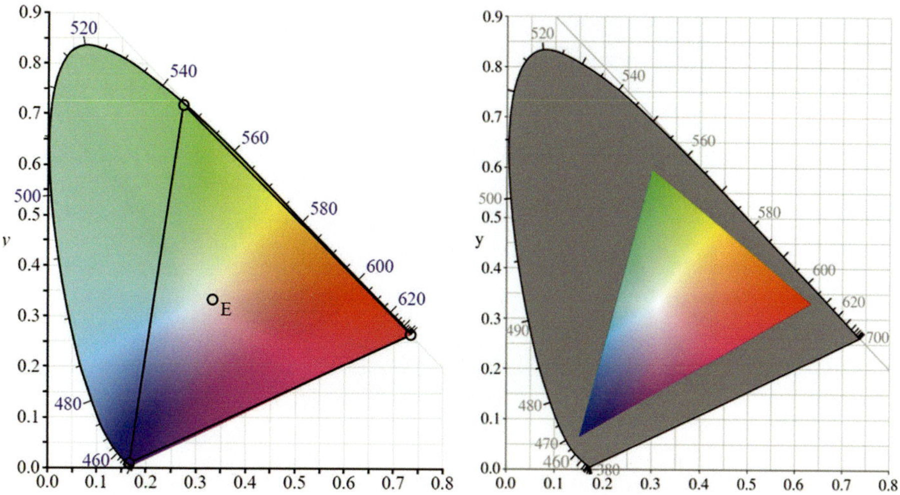


**Fig. 5.9** RGB and cathode-ray tube gamuts—by BenRG—own work, inspired by File: CIExy1931.png, Public Domain, https://commons.wikimedia.org/w/index.php?curid=7889718 and by Hankwang at English Wikipedia—transferred from en.wikipedia to Commons by Aboalbiss, Public Domain, https://commons.wikimedia.org/w/index.php?curid=8284581

In Chap. 4 we mentioned that different primary colors are employed in industry. Figure 4.50 illustrated the *RGB* scheme—defined by the primaries 700 nm red, 546.1 nm green, and 435.8 nm blue—as a triangle in CIE space. The triangle with these primary/corners is known as the *gamut* of colors for this scheme. Other choices for the primaries yield different gamuts. For example, the primary colors employed by a typical computer monitor are determined by its phosphors, and they span the triangle/gamut shown in Fig. 5.9.[2]

It is not surprising that these variations result in slight differences in the shades of colors that crop up in various media (including this book). Figure 5.10 displays a comparison of an image printed using an RGB versus a CYMK scheme.

[2] The website https://commons.wikimedia.org/wiki/File:CIE1931xy_gamut_comparison.svg displays gamuts for a number of different schemes employed in practice—including the printer's *SWOP CMYK* (cyan, magenta, yellow, black) scheme, whose gamut is pentagonal due to the "extra" black primary.

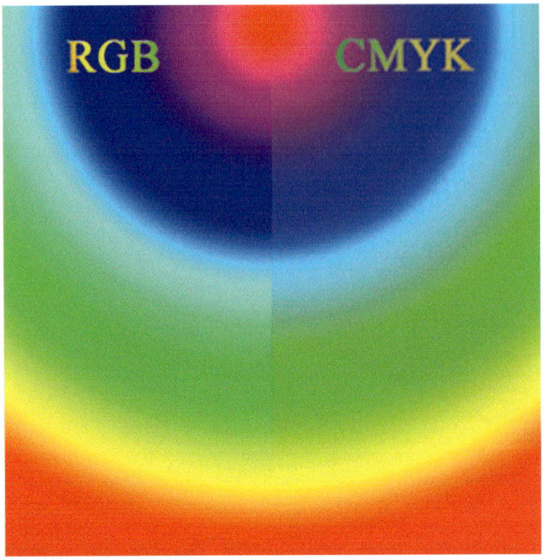


**Fig. 5.10** RGB versus CMYK representation—by RGB_CMYK_4.jpg: Annette Shacklettderivative work: Marluxia.Kyoshu (talk)—made in Photoshop. Original was also made in Photoshop in 2003. Public Domain, https://commons.wikimedia.org/w/index.php?curid=9396397

## 5.3 Pointillism: Georges Seurat

The 1963 award-winning musical "Sunday in the Park with George" by Steven Sondheim and James Lapine was inspired by **George Seurat**'s (1859–1891) best-known painting, *A Sunday Afternoon on the Island of La Grand Jette* (Fig. 5.11).

Seurat was one of the first painters to exploit the addition of colors by juxtaposition ("pointillism") on a grand scale. At the Art Institute of Chicago you can stand very close to the Grand Jette canvas and see the meticulous network of colored dots (Fig. 5.12).

Seurat studied books by color theorists describing how nearby colors affect the cones' responses to adjacent colors, and he formulated principles of color arrangements that he used in his paintings. If you enlarge Figs. 5.13 and 5.14 you'll observe his devotion to pointillism.

**Roy Lichtenstein** (1923–1997) carried pointillism to the extreme. He took the vulnerable point of newsprint comics—the necessity of using juxtaposed dabs of primary colors to depict objects—and made it his trademark. His dot patterns (originally developed by the printer/illustrator **Benjamin Day** in 1879) are perfect circles, and they are blatantly visible; their nonrandom echelon alignments are part of the artistic presentation. You will enjoy—and probably recognize—many examples of his work if you scan the web (https://www.wikiart.org/en/roy-lichtenstein/all-works#!#filterName:all-paintings-chronologically,resultType:masonry). Note particularly *Nurse*; it sold for $95,365,000 in auction at Christie's in Nov. 2015.

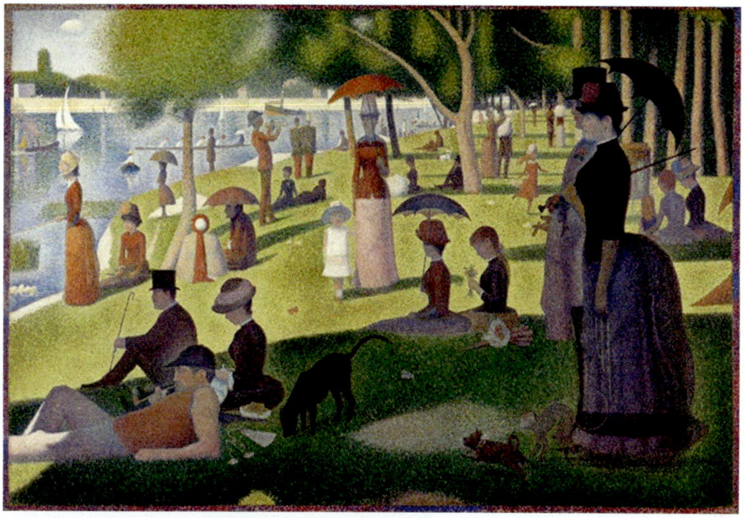

**Fig. 5.11** *La Grand Jette* (**Seurat**) Art Institute of Chicago, Public Domain, https://commons.wikimedia.org/w/index.php?curid=11500834

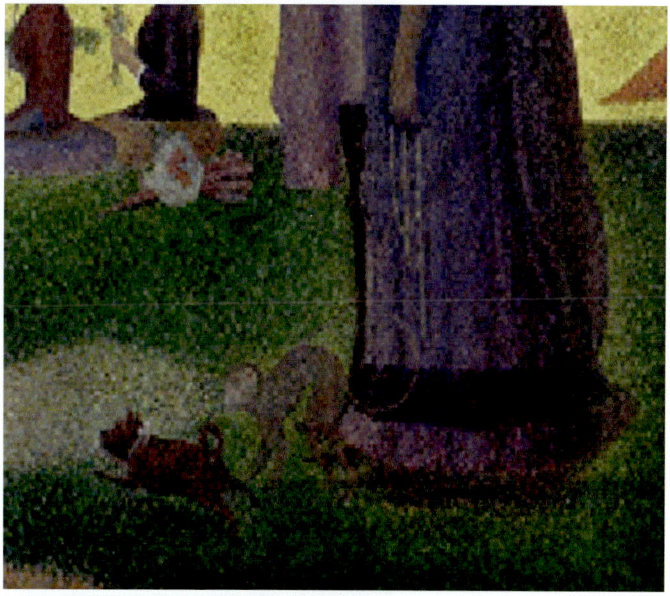

**Fig. 5.12** Detail (**Seurat**)—Art Institute of Chicago, Public Domain. https://commons.wikimedia.org/w/index.php?curid=11500834. Cropped and embellished from the original

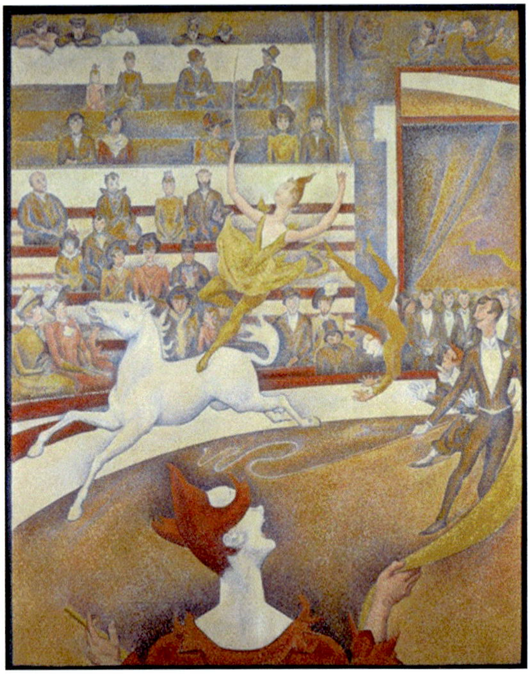

**Fig. 5.13** *The Circus* (**Seurat**)—by Georges Seurat—The Yorck Project (2002) 10.000 Meisterwerke der Malerei (DVD-ROM), distributed by DIRECTMEDIA Publishing GmbH. ISBN: 3936122202. Public Domain, https://commons.wikimedia.org/w/index.php?curid=158896

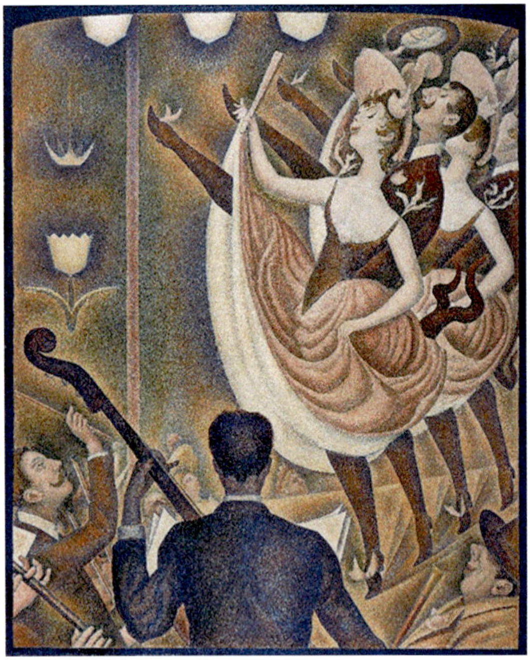

**Fig. 5.14** *Le* Chahut (**Seurat**)—Wikimedia Commons, File: Georges Seurat—Le Chahut—Google Art Project.jpg, Public Domain, https://commons.wikimedia.org/w/index.php?curid=44500865

## 5.4 The Munsell Color Tree

Engineers, you're not going to like this section. It's full of subjective, hazy, ill-defined, elusive-to-quantify nomenclature that will aggravate you. If you're like me, this probably kept you out the art elective courses all along. You see, painters had their own color classification schemes before the physics-based 1931 C.I.E. codes. They needed them; if an auto company manufactured a car door in Detroit and a fender in Toledo, it was impractical to bring the door to Toledo in order to match the colors.

In fact schemes for categorizing colors were conceived in the eighteenth century. They were based on three-degree-of-freedom systems, consistent with the trio of primary colors on Maxwell's spinning disks and the trinity of cone types in the eye. One parameter in all these schemes was the basic color, or "hue", of the observed light: red, violet, yellow, gold, …; twenty hue values are displayed in the color wheel in Fig. 5.15.

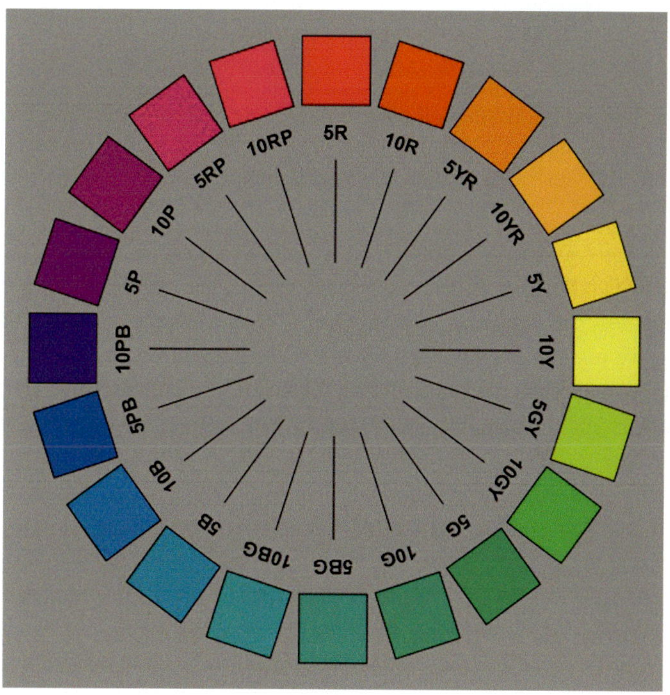


**Fig. 5.15** Munsell color wheel—by Thenoizz—Own work, CC BY 3.0, https://commons.wikimedia.org/w/index.php?curid=5216531

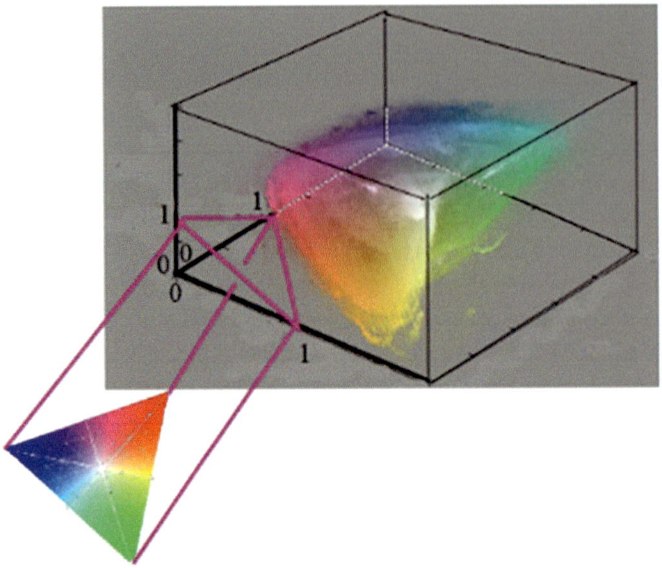


**Fig. 5.16** (Fig. 4.48 repeated) Maxwell color triangle—by Niabot—own work, CC0, https://commons.wikimedia.org/w/index.php?curid=12407052; and by Polina Tishina—Polina Tishina, Public Domain, https://commons.wikimedia.org/w/index.php?curid=9816956. Cropped and embellished from the original

But the other two degrees of freedom constituted a chaos of descriptors among artists, including "lightness", "brightness", "saturation", "tone", "shade", and the like.[3] Remember that when we mathematically reduced the 3D red/green/blue octant of colors to a 2D triangle earlier, we sacrificed recognition of these fine distinctions (Fig. 5.16).

In the early twentieth century Albert Henry Munsell (1858–1918), a Massachusetts painter and art teacher, organized a numerical 3D color classification scheme that survives to this day. After consultations with many artists, photographers, engineers, and physicists he was able to achieve a consensus that the remaining two variables should be *value* and *chroma*.

The vertical scale in Fig. 5.17 displays a sample of the *values* that can be taken by forty different hues.

Samples of the *chroma* measure for five hues are displayed horizontally in Fig. 5.18.

Experts tell me that value indicates the brightness of the color, from black to white; while chroma represents its purity, from "washed-out" to "saturation".

[3] Wikipedia, the free Encyclopedia, valiantly strives to provide a lexicon at https://en.wikipedia.org/wiki/HSL_and_HSV.

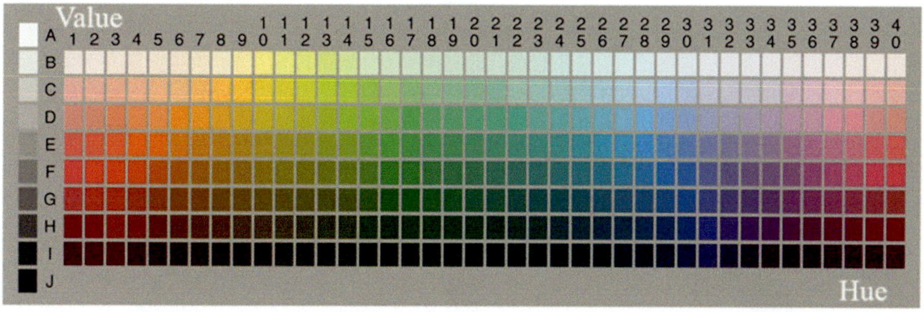


**Fig. 5.17** Hue and Value—by Mikael Vejdemo-Johansson, Susanne Vejdemo, Carl-Henrik Ek—https://journals.plos.org/plosone/article?id=https://doi.org/10.1371/journal.pone.0089184, CC BY 4.0, https://commons.wikimedia.org/w/index.php?curid=97226766. Embellished from the original

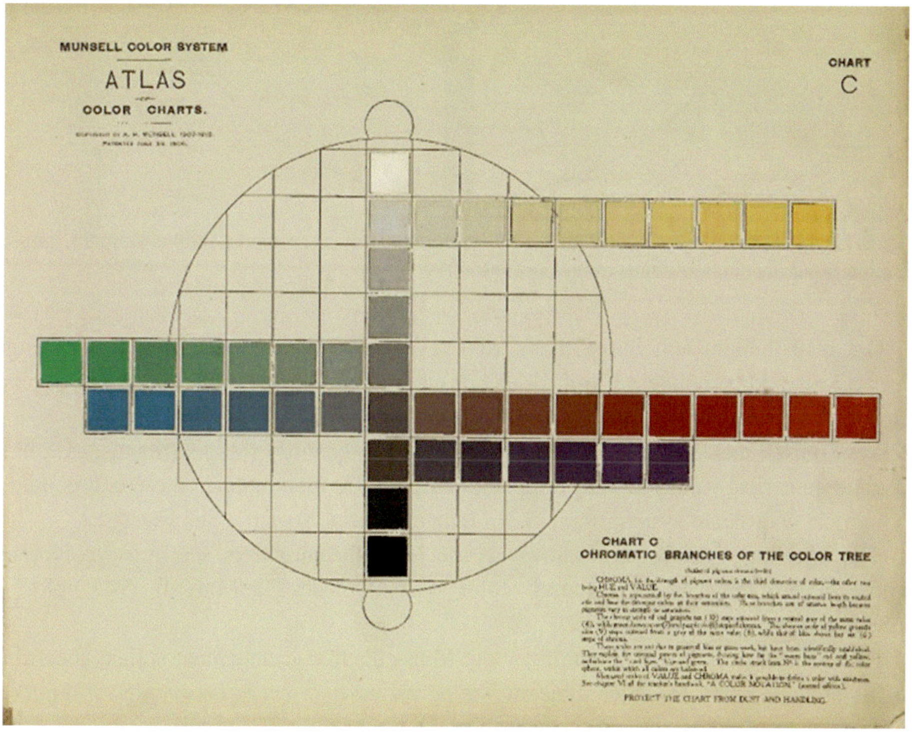


**Fig. 5.18** Hue and Chroma—by Albert H. Munsell; Page 13 of *Atlas of the Munsell Color System*, 1915 edition; https://library.si.edu/digital-library/book/atlasmunsellcol00muns. Public Domain https://en.m.wikipedia.org/wiki/File:Atlas_of_the_Munsell_Color_System_page_13.jpg

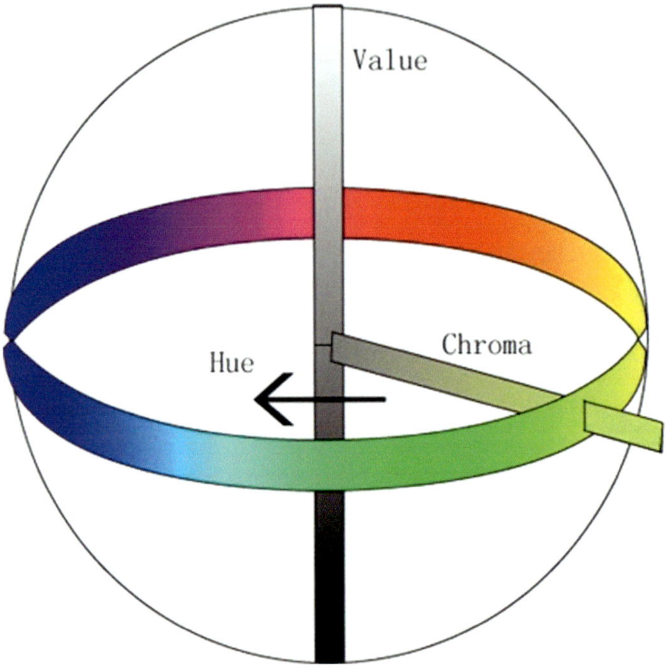


**Fig. 5.19** HVC cylindrical coordinates—by Tsiaojian_Lee—own work, Public Domain, https://commons.wikimedia.org/w/index.php?curid=864039

Got it? I thought not. Sorry, that's all I've got. I'm not an artist, I'm just an engineer who would be helpless if not for electronic hue/saturation/brightness translators like Florida State University's Molecular Expressions Website interactive converter.[4]

Yet Munsell tackled the thankless task of trying to quantify these attributes, assigning discrete numerical values to hue, value, and chroma. He then mapped these colors onto a cylindrical coordinate system $(\theta, z, r) = (Hue, Value, Chroma)$. See Fig. 5.19.

The resulting array became known as the Munsell color tree; see https://commons.wikimedia.org/wiki/Category:Munsell_color_system#/media/File:Munsell_1943_color_solid_cylindrical_coordinates.png.

Perhaps the most remarkable thing about Munsell's tree arrangement is that, according to his polls of many observer-subjects, *the perceived differences between each adjacent pair of colors is uniform.* Here adjacent means left-and-right *and* above-and-below. Standardization of color measurements is plagued with difficulties like variations in lighting sources, degrees of color blindness in the subjects, the particular area of the retina that is illuminated, personal tastes, etc. However, Munsell's HVC classification system is just as well-defined as the C.I.E. or RGB systems.

[4] http://micro.magnet.fsu.edu/primer/java/scienceopticsu/colorstructure/.

The Hue/Value/Chroma parametrization is employed today by industries classifying skin color, soil color, tooth color, and beer color (!). Another system, parametrizing the Hue/Saturation/Lightness attributes, has been found advantageous in broadcasting analog television (although the signal is converted for RGB display on the screen, Fig. 5.3). Other issues such as computational convenience and computer vision needs have birthed Hue/Saturation/Value systems and Hue/Saturation/Intensity systems.[5]

In addition, Munsell was able to formulate principles for achieving aesthetic color harmony in terms of his HVC coordinates! For example, if two colors from opposite sides of the HVC $z$-axis appear in the same painting, Munsell recommends that the areas occupied by the two colors be inversely proportional to the products of their values and chroma.[6]

I'm going to close my book by describing the works of three artists who I think are particularly notable in the way they manipulate colors: Modigliani for red, Cezanne and van Gogh for green.

## 5.5 Amedeo Modigliani

Believe it or not, the first thing you'll notice about the painting in Fig. 5.20 when you walk into Gallery 913 of the Modern and Contemporary Art Department of New York's Metropolitan Museum of Art is the color! Unfortunately, neither your computer screen nor the printed reproductions of *Reclining Nude* can capture the earthiness of these bawdy crimsons. **Amadeo Modigliani** (1884–1920) conjured up the most sensual shades of red for the couch and the model's skin; coupled with the erotic nature of the painting, viewing the original never fails to bring an embarrassing flush of excitement to this engineer.

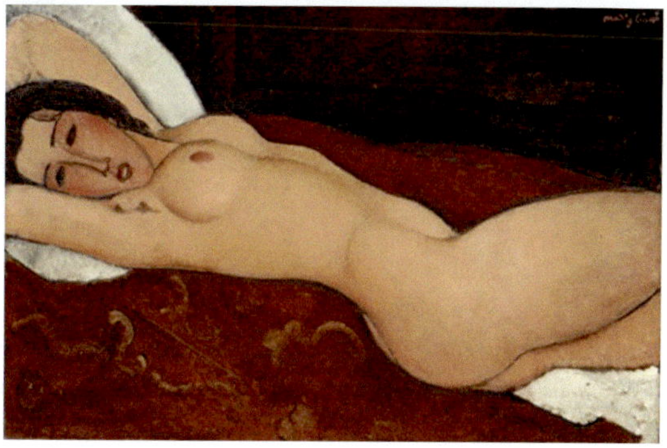

**Fig. 5.20** *Reclining Nude* (**Modigliani**)—Metropolitan Museum of Art, Public Domain, https://commons.wikimedia.org/w/index.php?curid=36219851

[5] https://en.wikipedia.org/wiki/hsl_and_hsv.

[6] http://munsell.com/color-blog/grammar-of-color-munsell-book-online/; https://www.yumpu.com/en/document/view/10535340/a-historical-review-of-proposals-for-color-harmony.

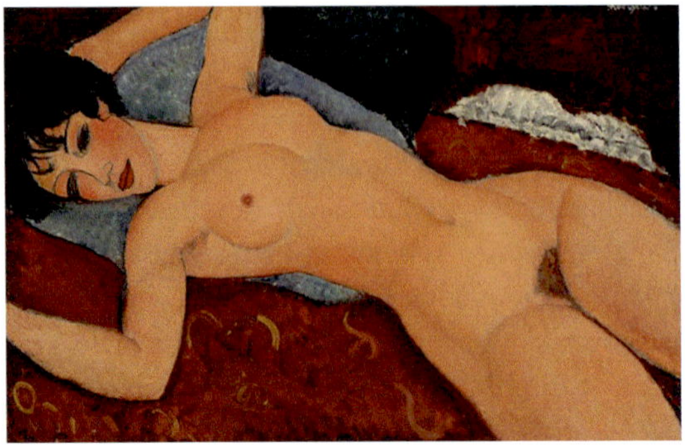

**Fig. 5.21** *Nu Couche* (Red Nude) (**Modigliani**)—Christie's sale, New York, 9 November 2015, Public Domain, https://commons.wikimedia.org/w/index.php?curid=44456944

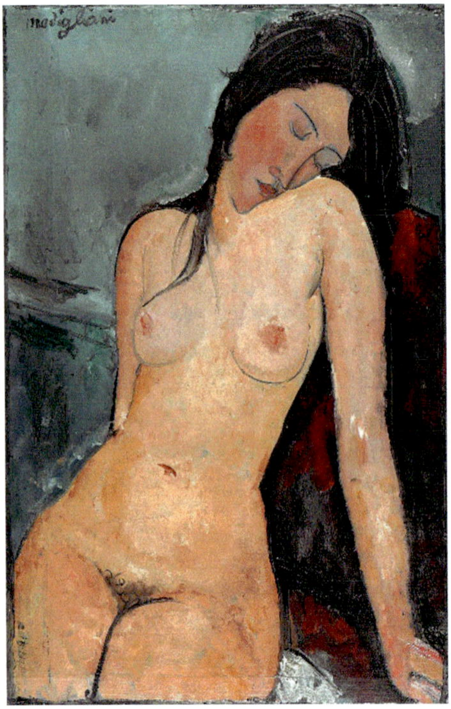

**Fig. 5.22** *Seated nude* (**Modigliani**)—1. The Yorck Project (2002) 10.000 Meisterwerke der Malerei (DVD-ROM), distributed by DIRECTMEDIA Publishing GmbH. ISBN: 3936122202. 2./3. Unknown source. 4. Courtauld Institute of Art, online collection, Public Domain, https://commons.wikimedia.org/w/index.php?curid=155804

Modigliani was notorious for his sensual nudes. He painted around 20 of them in the winter of 1917, and the police demanded the closure of his first and only one-man show at the Galerie Berthe Weill in Paris. But I think you'll agree *Reclining Nude* was not his most beautiful model.

*Red Nude* (Fig. 5.21) sold at auction in November of 2015 for $170,405,000, at Christie's in New York. It had rarely been exhibited in the previous 40 years, but the

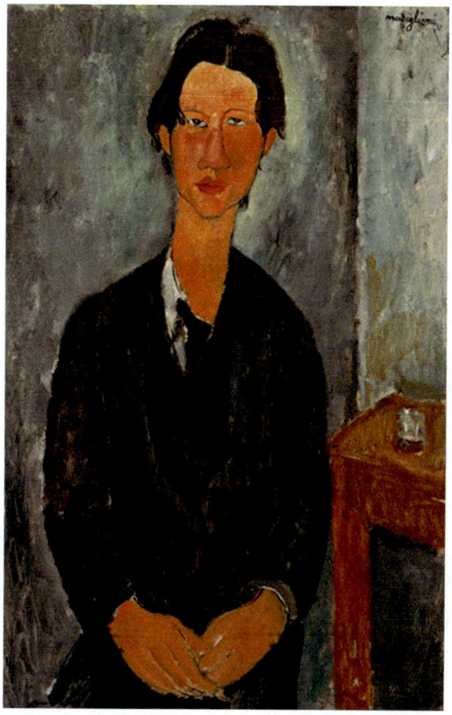

**Fig. 5.23** Portrait of Chaim Soutine (**Modigliani**)—http://www.nga.gov/content/ngaweb/Collection/art-object-page.46522.html, Public Domain, https://commons.wikimedia.org/w/index.php?curid=34866518

image is well-known from art books, highball glasses, college dorm posters, etc.[7] The same couch features in many others of Modigliani's paintings,[8] but its color is not so marvelously coupled with the model's skin (Fig. 5.22).

Modi's portraits of his drinking buddies are remarkable. His style—the graceful elongated necks, the shaded eyes (enlarge the images to appreciate this)—permeates all of his work (Figs. 5.23, 5.24 and 5.25).

Tragically, Modigliani's paintings never made very much money during his lifetime. At one point he offered the entire collection in his Paris studio for $300, and had no takers. After a career of poverty, drinking, drug abuse, and fornication, he died of tubercular meningitis at 35.

[7] I saw *Red Nude* at the auction—the first and only time I saw the original—and its coloring even surpasses that of *Reclining Nude.*

[8] (If only it could talk!)

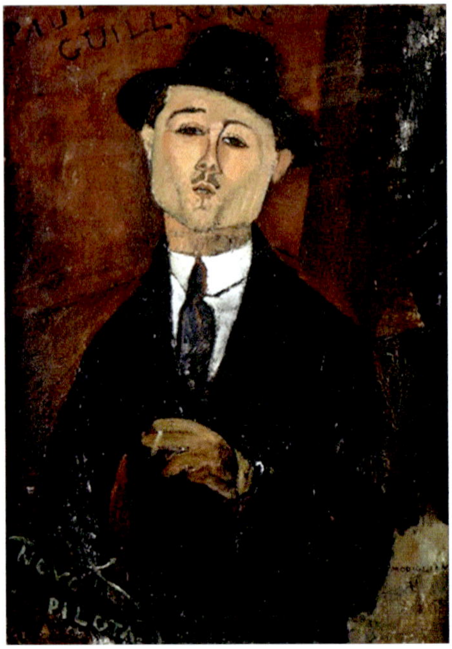

**Fig. 5.24** Portrait of Paul Guillaume (**Modigliani**)—own work, user: Rlbberlin, Public Domain, https://commons.wikimedia.org/w/index.php?curid=2471508

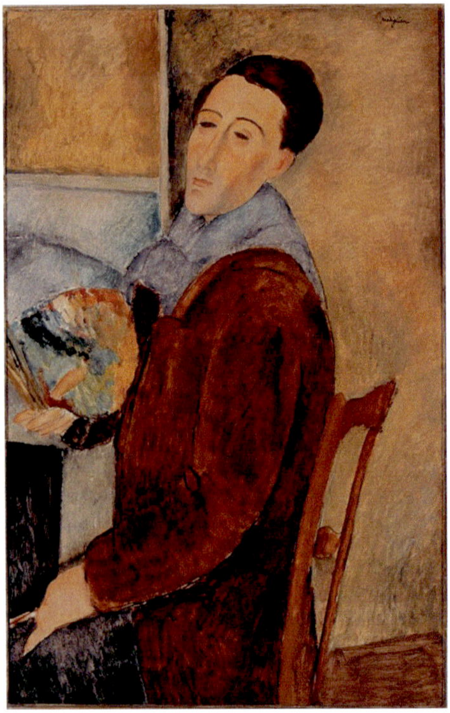

**Fig. 5.25** Self portrait (**Modigliani**)—http://www.mac.usp.br, Public Domain, https://commons.wikimedia.org/w/index.php?curid=2172374

## 5.6 Paul Cezanne

I may be revealing a personal preference here, but it seems to me that the greens in **Cezanne**'s (1839–1906) paintings are singularly beautiful. Here are a few samples (Figs. 5.26, 5.27, 5.28 and 5.29).

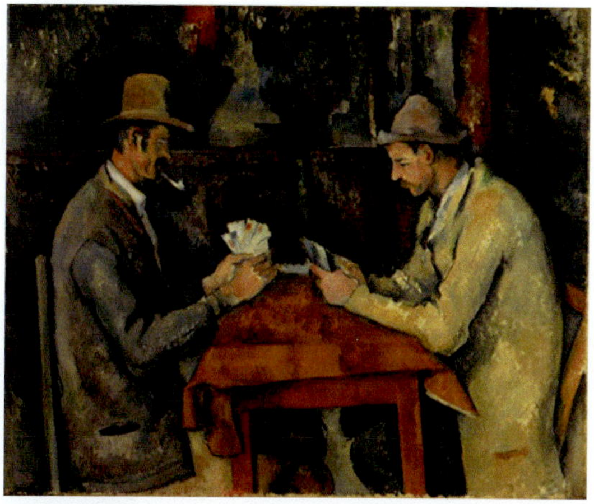

**Fig. 5.26** *The Card Players* (**Cézanne**)—Courtauld Institute of Art, Public Domain, https://commons.wikimedia.org/w/index.php?curid=31084348

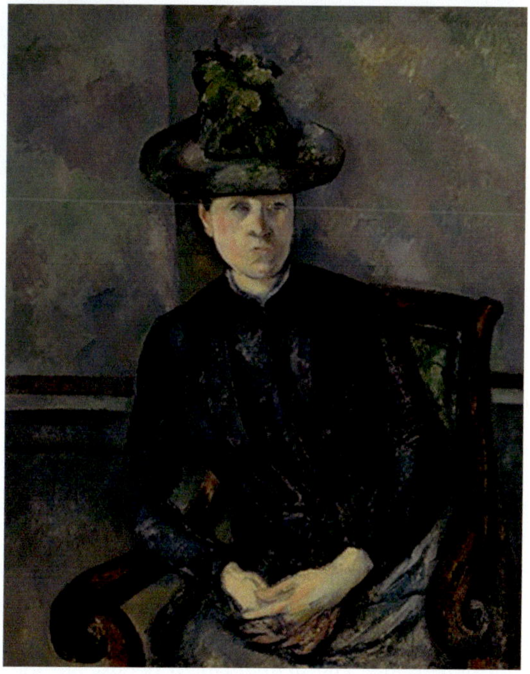

**Fig. 5.27** *Woman in a green hat* (**Cézanne**)—https://collection.barnesfoundation.org/objects/6953/details, Public Domain, https://commons.wikimedia.org/w/index.php?curid=67856062

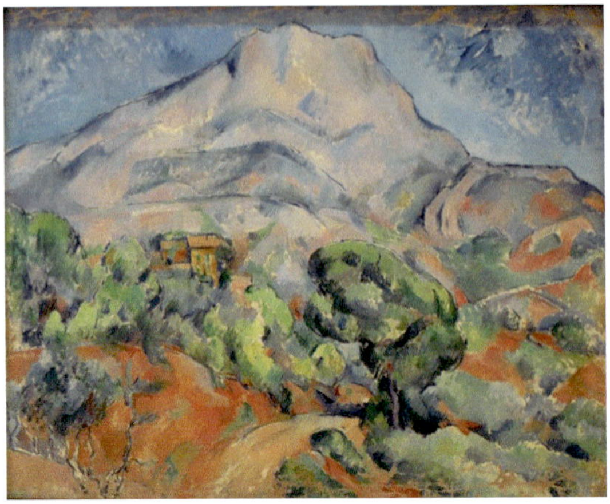

**Fig. 5.28** *Mont Sainte-Victoire seen from Bellevue* (**Cézanne**)—own work, Public Domain, https://commons.wikimedia.org/w/index.php?curid=149179

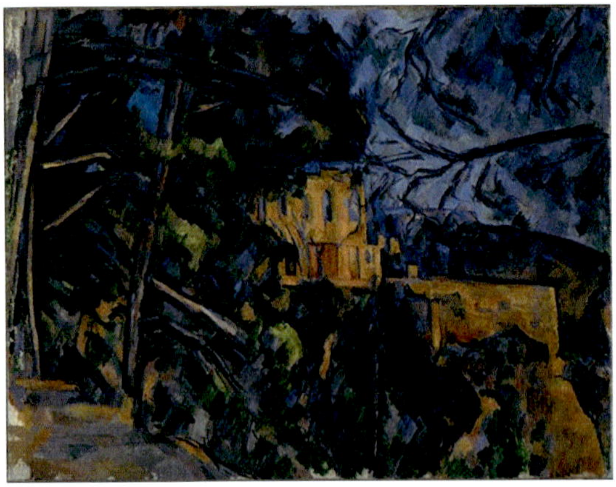

**Fig. 5.29** *Chateau*-Noir (Cezanne)—by Erwin Jurschitza; Public Domain, https://commons.wikimedia.org/w/index.php?curid=148984

But there's another feature in Cezanne's works that has intrigued art historians. First of all, note how Cezanne ignores the principles of perspective as we discussed them in Chap. 3 (Figs. 5.30 and 5.31).

Indeed, a popular joke of the time went: "Cezanne was a wonderful artist, but I wouldn't want him to be my carpenter." Yet Cezanne does convey a sense of depth to the work. How is that?

In Chap. 3 we mentioned (briefly) that the rules of so-called "linear" perspective only hold, rigorously, for viewing in *one eye*. The left eye will extrapolate vanishing points to slightly different locations than the right eye. "Binocular" perspective was analyzed extensively in a 1950 paper (and subsequent books) by Rudolph K. Luneberg (1903–1949) ("The Metric of Binocular Visual Space", *Journal of the Optical Society of America,* vol. 40 #10, 627–642). Luneberg argued that the geometry of binocular vision is not Euclidean, but hyperbolic. The mathematical details are too complicated for our present study, but

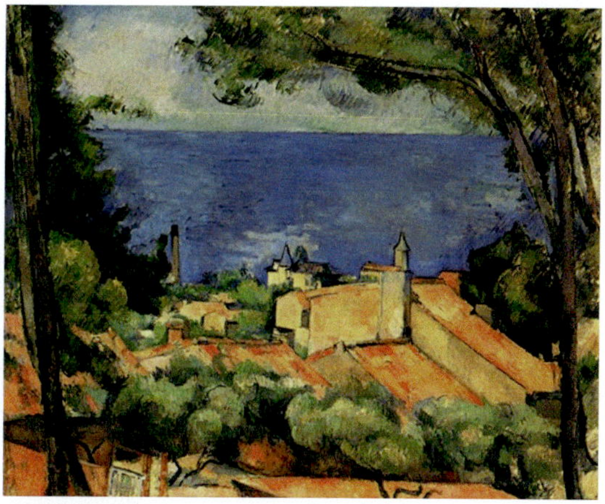

**Fig. 5.30** *L'Estaque* (**Cézanne**)—The Yorck Project (2002) 10.000 Meisterwerke der Malerei (DVD-ROM), distributed by DIRECTMEDIA Publishing GmbH. ISBN: 3936122202. Public Domain, https://commons.wikimedia.org/w/index.php?curid=149059

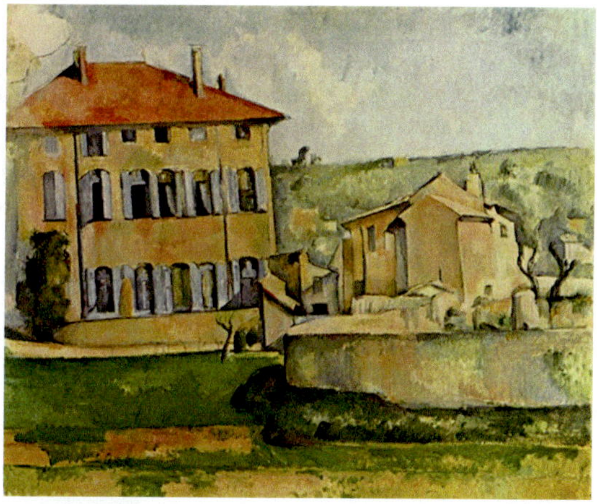

**Fig. 5.31** *Jas de Bouffan* (**Cézanne**)—The Yorck Project (2002) 10.000 Meisterwerke der Malerei (DVD-ROM), distributed by DIRECTMEDIA Publishing GmbH. ISBN: 3936122202. Public Domain, https://commons.wikimedia.org/w/index.php?curid=149048

art experts have noted how Cezanne seemed to be aware of this; note that his trees in Fig. 5.32 take the shapes of hyperbolas in this painting.

Another artifact of binocular vision is that two images of a scene, from slightly different viewpoints, provide depth. Close your left eye and hold your right hand in front of your face, as shown in Fig. 5.33.

Twist your hand until you can no longer see (in your right eye) the little finger of the hand; it is blocked by the first fingers. Now close the right eye and open the left; the new perspective provides a view of the little finger, and hence adds depth to the right eye's image.

**Exercise** Close one eye, and ask two friends to stand facing each other about 10 feet away from you, with their arms slightly extended towards each other. Verbally direct one of them

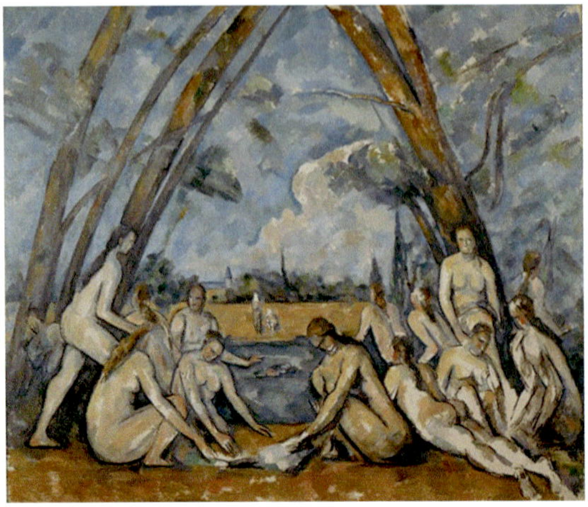

**Fig. 5.32** *Les Grandes Baigneuses* (The Large Bathers) (Cézanne—owGszazN_Evyvw at Google Cultural Institute maximum zoom level), Public Domain, https://commons.wikimedia.org/w/index.php?curid=21885960

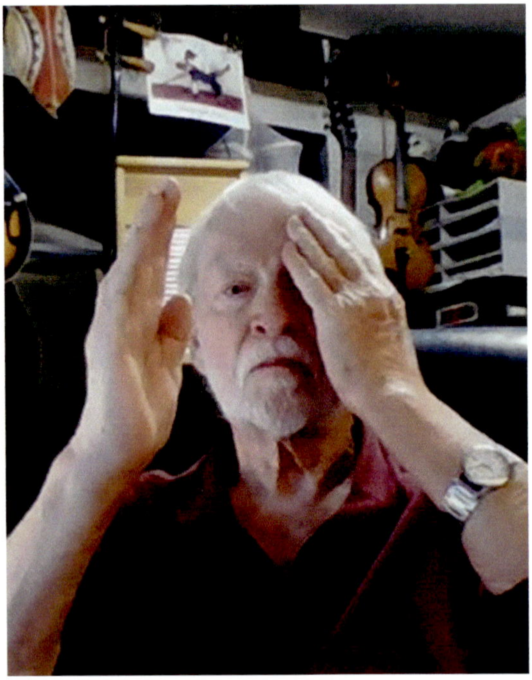

**Fig. 5.33** Binocular vision demonstration

to move toward or away from you until you judge they are at equal distances from you. Then open the other eye. If you're like most people, you will be surprised to see that you have misaligned them by about one foot. This is a dramatic demonstration of the role of binocular vision in depth perception.

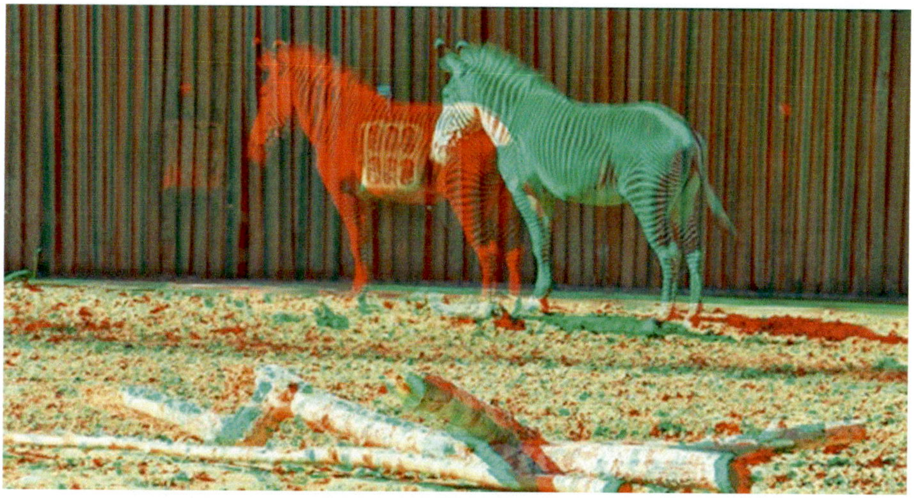

**Fig. 5.34** IMG_6106a1-Anaglyph Photo/3D—by relaxed now. CC BY 2.0 Deed https://www.flickr.com/photos/relaxednow/48919228097/

The "three-dimensional" comics produced in the early 1950s relied on this effect. A cartoon was drawn twice from slightly different angles, one drawing in red and the other in blue/green. When viewed through glasses which blocked red light in one eye and green in the other, a depth dimension is definitely added to the so-called "anaglyph" (Figs. 5.34, 5.35 and 5.36).

The original 3D comic (Mighty Mouse) can be seen at http://captainvideossecretsanctum.blogspot.com/2012/02/3-d-three-dimensional-comics-mighty.html.

**Exercise** Rig up a set of 3D glasses using two pieces of red and green cellophane or transparent plastic and view Figs. 5.34 and 5.35.

Some art experts have conjectured that Cezanne was experimenting with this trick to give the impression of depth in some of his paintings (recall Fig. 5.32). Picasso and Matisse called Cezanne the "Father of Cubism," inspiring them to carry (to extremes!) the notion of two (or more) simultaneous views of an object in one painting: see https://en.wikipedia.org/w/index.php?curid=59488370 and https://en.wikipedia.org/wiki/List_of_Picasso_artworks_1911%E2%80%931920.

Finally, note that the 3D aspect of paintings using linear perspective should be better when they are viewed with one eye than with two; while paintings with binocular perspective should behave just the opposite. Try this out on daVinci's *Last Supper* (Fig. 2.82) and Cezanne's *Large Bathers* (Fig. 5.32). What do you think?

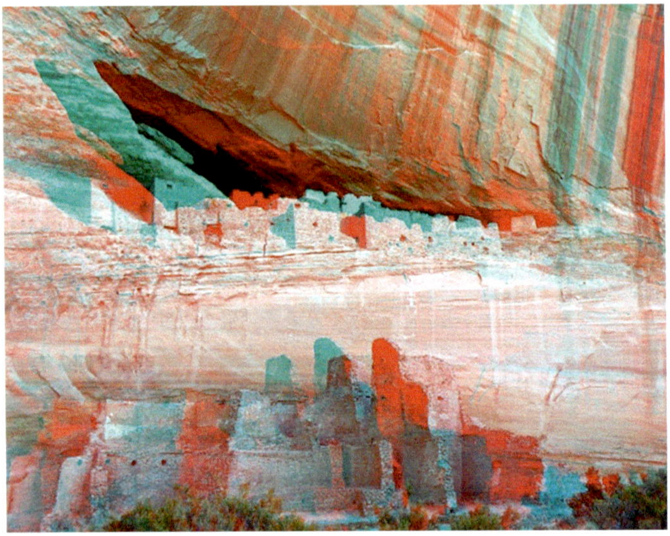

Fig. 5.35 White House ruins—by relaxed now—ABD_3218—anaglyph photo/3D, CC BY 2.0, https://commons.wikimedia.org/w/index.php?curid=97531047

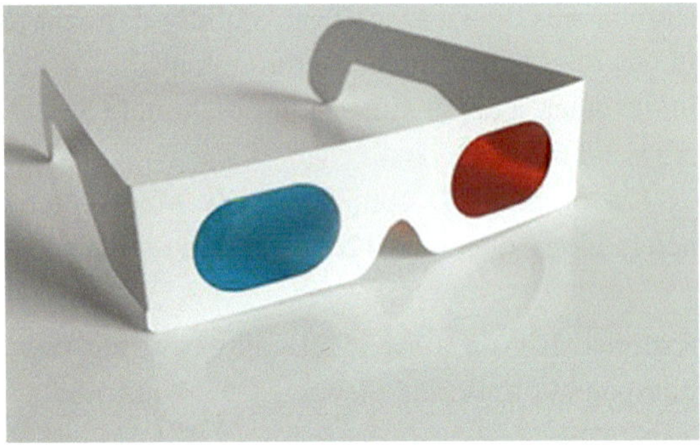

Fig. 5.36 3D glasses—by xenmate CC BY 2.0 DEED https://www.flickr.com/photos/xenmate/6673633381

## 5.7 Vincent Willem van Gogh

Like Paul Cezanne, **van Gogh** (1853–1890) was a master of greens. He did not start painting until his late twenties. His early works were quite devoid of color, but as he developed his style they evolved into the frenzied profusions that we associate with this troubled genius (Figs. 5.37, 5.38, 5.39, 5.40, 5.41, 5.42, 5.43 and 5.44).

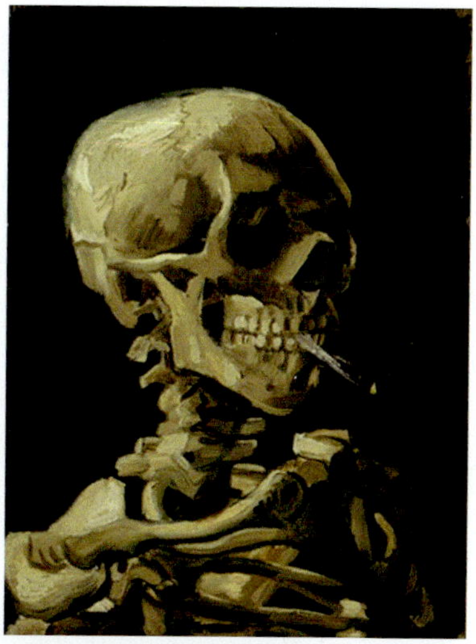

**Fig. 5.37** *Skull of a skeleton with burning cigarette* (**van Gogh**)—Museum Page (the image was stitched from tiles—for assembly method used see The Potato Eaters; https://www.vangoghmuseum.nl/nl?page=1628&collection=619&lang=en), Public Domain, https://commons.wikimedia.org/w/index.php?curid=22030423

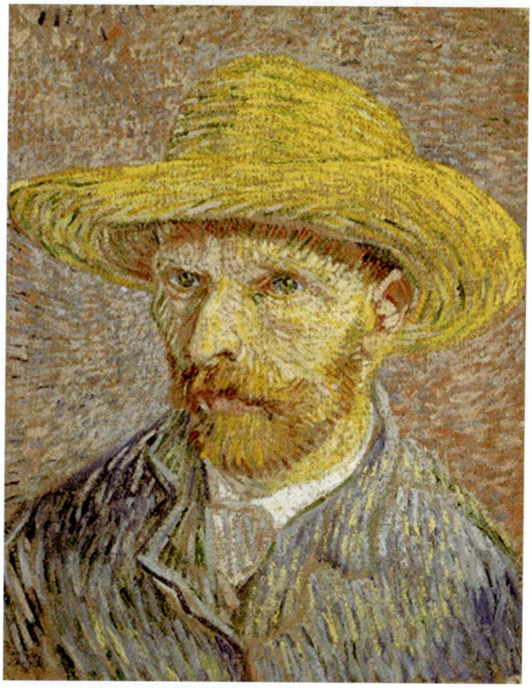

**Fig. 5.38** *Self-portrait with straw hat* (**van Gogh**)—art database, Public Domain, https://commons.wikimedia.org/w/index.php?curid=6491390

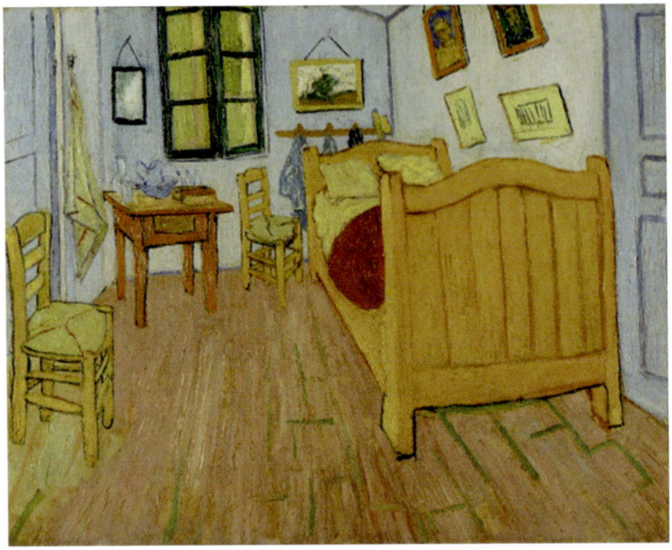

**Fig. 5.39** *Bedroom in Arles* (**van Gogh**)—by KwF-AdF1REQl6w—Google Arts and Culture, Public Domain, https://commons.wikimedia.org/w/index.php?curid=22554513

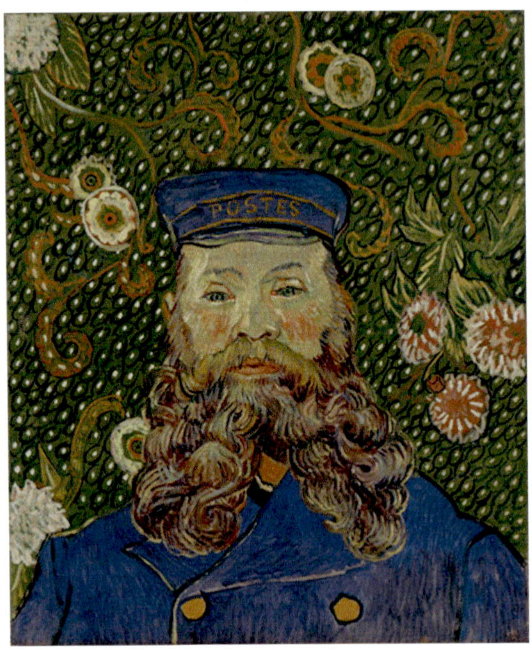


**Fig. 5.40** *Portrait of the Postman Joseph Roulin* (**van Gogh**)—dgHM8pgTjWF6Cg—Google Arts and Culture, Public Domain, https://commons.wikimedia.org/w/index.php?curid=13357994

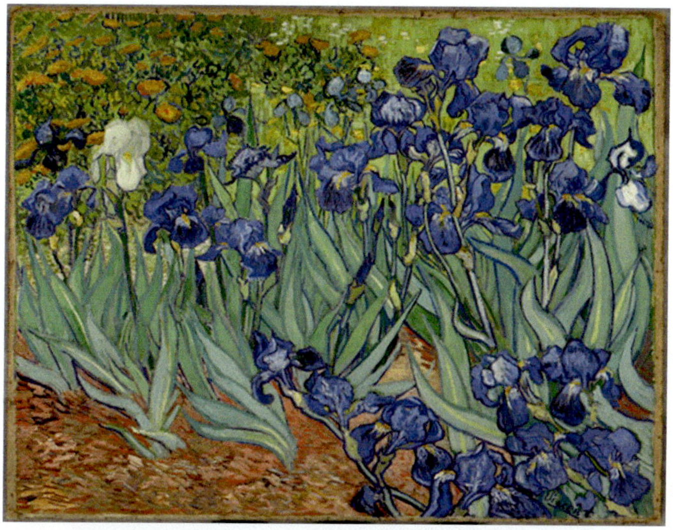

**Fig. 5.41** *Irises* (**van Gogh**)—DgFVFAJo_30MeQ at Google Cultural Institute maximum zoom level, Public Domain, https://commons.wikimedia.org/w/index.php?curid=22175016

**Fig. 5.42** *Portrait of Dr. Gachet* (**van Gogh**)—OQE_w8xp1TtbSw at Google Cultural Institute maximum zoom level, Public Domain, https://commons.wikimedia.org/w/index.php?curid=21880316

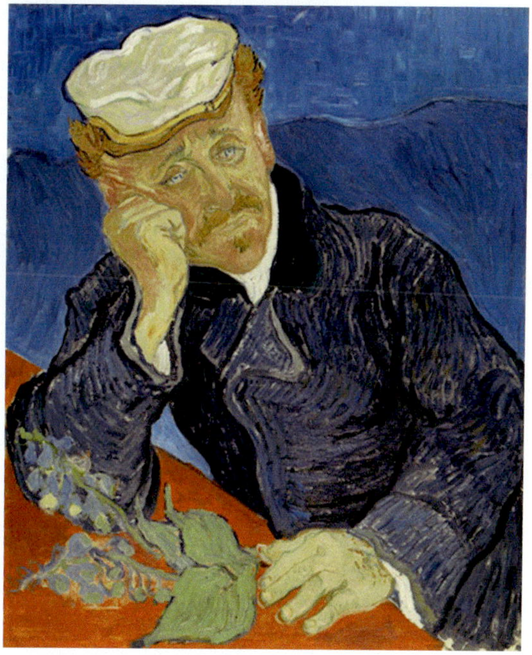

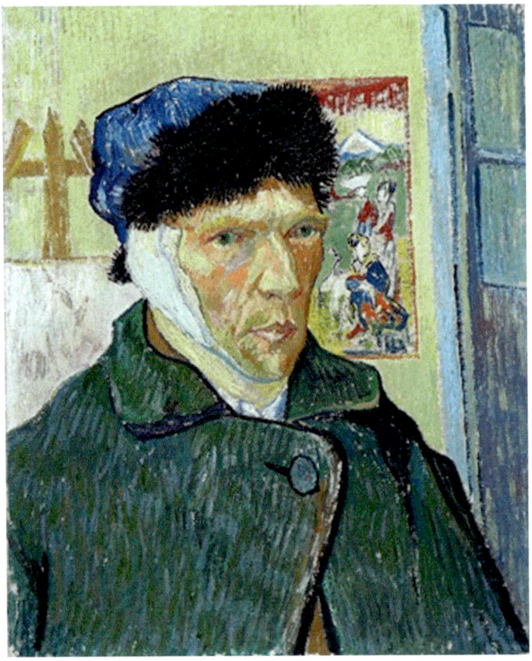

**Fig. 5.43** *Self-portrait with bandaged ear* (**van Gogh**)—The Courtauld Institute of Art, Public Domain, https://commons.wikimedia.org/w/index.php?curid=29334036

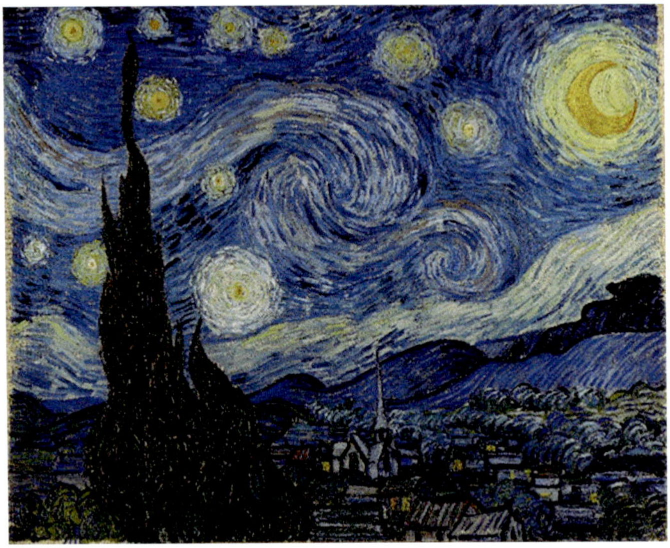

**Fig. 5.44** *Starry night* (**van Gogh**)—bgEuwDxel93-Pg—Google Arts and Culture, Public Domain, https://commons.wikimedia.org/w/index.php?curid=25498286

These masterpieces reveal van Gogh's prowess with color juxtaposition; you don't see those individual colors in the sky in *Starry Night*, or in a jacket in *Self Portrait with Straw Hat*. The latter work also suggests some experimenting with pointillism. Some art historians attribute the anarchy in his later works with the artist's descent into mental illness; despite his treatment under Dr. Gachet, he died from an allegedly self-inflicted gunshot wound at age 37.

*Portrait of Dr. Gachet* was sold for $82.5 million in 1990.

How's that for green, dear readers?

∂

# Postface

My STEM (excuse me; **STEAM**) friends, I began work on this book in 2007, and several things have changed over the years. The biggest jolt has been in the face value of the most in-demand pieces of artwork. The figures I cite in the text, in the hope of stimulating some excitement in your technology-transfused bones, have been surpassed so far as to render them laughable.

When I went to Christie's auction in New York in November 2015, for the first and only time in my life that I would see the original Nu Couche by Modigliani (Fig. 5.21), I was thrilled that my beloved fantasy garnered $170,405,000—the second-highest amount ever paid for a painting. (Well, to be honest, Picasso's Les Femmes d'Alger had sold for $179,000,000 in May of that year. But still!).

Turn the clock ahead to 2024.

Da Vinci's Salvator Mundi[1] sold for $450,500,000 (!)

De Kooning's Interchangeinterchange[2] sold for $300,000,000. Cezanne's Card Players (Fig. 5.26) sold for $250,000,000.

Gauguin's Nafea Faa Ipoipo[3] sold for $210,000,000. Pollock's Number 17A[4] sold for $200,000,000.

Rembrandt's Standard Bearer[5] sold for $197,900,000. Warhol's Shot Marilyn[6] sold for $195,000,000.

---

[1] https://en.wikipedia.org/wiki/Salvator_Mundi_(Leonardo).
[2] https://en.wikipedia.org/wiki/Interchange_(de_Kooning).
[3] https://en.wikipedia.org/wiki/When_Will_You_Marry%3F.
[4] https://en.wikipedia.org/wiki/Number_17A.
[5] https://en.wikipedia.org/wiki/The_Standard_Bearer_(Rembrandt,_1636).
[6] https://en.wikipedia.org/wiki/Shot_Marilyns.



A. D. Snider, *From STEM to STEAM*, Synthesis Lectures on Engineering, Science, and Technology, https://doi.org/10.1007/978-3-031-57316-3

Rothko's No. 6[7] (Violet, Green, and Red) sold for $186,000,000. Klimt's Wasserschlangen II[8] sold for $183,000,000.

Rembrandt's Portraits of Maerten Soolmans and Oopjen[9] sold for $180,000,000. Then come Picasso and Modigliani.

And of course the Louvre would never sell Da Vinci's Mona Lisa (Fig. 2.81), but its value is estimated at $970,000,000.

Maybe it's time to get out of science, technology, engineering, and mathematics.

7 https://en.wikipedia.org/wiki/No._6_(Violet,_Green_and_Red).

8 https://en.wikipedia.org/wiki/Water_Serpents_II.

9 https://en.wikipedia.org/wiki/Pendant_portraits_of_Maerten_Soolmans_and_Oopjen_Coppit.

MIX
Papier aus verantwortungsvollen Quellen
Paper from responsible sources
FSC® C105338

If you have any concerns about our products,
you can contact us on
**ProductSafety@springernature.com**

In case Publisher is established outside the EU,
the EU authorized representative is:
**Springer Nature Customer Service Center GmbH**
**Europaplatz 3, 69115 Heidelberg, Germany**

Printed by Libri Plureos GmbH
in Hamburg, Germany